MW00686215

DATE DUE

Expectations and the Food Industry

The Impact of Color and Appearance

Expectations and the Food Industry

The Impact of Color and Appearance

JOHN B. HUTCHINGS

Chartered Physicist
Fellow of the Institute of Physics
Fellow of the Institute of Food Science & Technology
Bedford, England

Kluwer Academic / Plenum Publishers
New York, Boston, Dordrecht, London, Moscow

ISBN 0-306-47291-0

©2003 Kluwer Academic / Plenum Publishers, New York
233 Spring Street, New York, New York 10013

http://www.wkap.com

10 9 8 7 6 5 4 3 2 1

A C.I.P. record for this book is available from the Library of Congress

To Jenny, Dickon, Charles, James,
Matthew, Silvia, Evelyn, and Vivien

Preface

We purchase an object or enter a scene not for their own sake but for the expectations we have of them. When we purchase an orange we do so in the expectation that it will quench our thirst or that it will taste good, or that it will make us healthy. On the other hand, our orange is so perfect looking (because it has been dosed with insecticide and herbicide) and shiny (because it has been coated with wax) that if we do not wash it thoroughly before eating we will eat it in the expectation that it will poison us.

The activity of the moment is pursued not only for duty or immediate pleasure, but also with the dread, excitement, or merely boredom of that which lies ahead. This applies whether we have a plate of food in front of us, we are entering a room, shopping, at work or play, or merely doing the washing up. We are continually experiencing expectations, most of them subconsciously. However, all lead to motivation and state of mind. Joy or disappointment results from the fulfilment or otherwise of prior expectations. In other words, the stimulus provided by the total appearance of an object or scene engenders expectations of the outcome of our involvement with the object or event.

Throughout the food chain, expectations are at the heart of quality judgements and prices are generated by expectation. Whether the vegetable crop is suitable for the high-end fresh frozen product or appropriate merely for inclusion in the heavily sauced spicy meal there are differences in expectation and value. The specimen strawberry that has a good enough appearance to be exhibited on the top of the dessert increases expectations of the whole dish.

The environment contributes significantly to the expectations we have of the food we are about to purchase. On entering the restaurant we judge the flavor of the space. Factors decreasing expectations include the waiter's dirty fingernails, the damaged food pack in the store, and the dirty tablecloth in the restaurant. Expectations arise from our visually perceived view, that is, the total appearance of the world around us. We deduce from the total appearance of the food in front of us whether it will harm us or be good for us. Similarly, we deduce from the total appearance of the restaurant or store whether foods associated with them are likely to harm us or be good for us.

Different people have different expectations from the same visual stimulus. The designer may see his or her store or restaurant as a work of art; those using the space may see it as an inconvenience, a place to be avoided. The designer designs and walks away; the user is left to manage the space in the best way he or she can. The effect of time, adaptation, wear and tear converts the designers' pristine creations into a real lived-in space. Many books are written with the aim of encouraging the designer and encouraging the exploitation and manipulation of the customer inhabiting the space, but little is written about the customers' perceptions, images, understanding, and expectations of the space. For a lifetime I have been associated with the food business, but this book records the point of view of a customer in whom expectations have been generated. Many happy, as well as unhappy, hours have been spent in noting what others have had in store for me in many daily personal food involvements and environments.

This book tackles expectations and how they arise, expectations associated with strangers involved in the food business, with the store, business façades, advertisements, and packaging, as well as expectations engendered in the restaurant and of the food itself. A holistic approach has been taken as total appearance images and expectations are critical in both separate and interlinking ways to all aspects of food research and development, production, marketing, sales, and preparation, as well as consumption. Above all they are critical to each individual customer whether he or she is in the kitchen, store, restaurant, or pub.

This is the first book to deal with the theory of appearance in toto, from color and total appearance to the creation and workings of expectations. It seeks to help those in all areas of the food industry who contribute to the visual environment experienced by the customer. These include architects, store designers, and food producers, whether they be banquet chef or manufacturer, as well as those in advertising and packaging or having responsibility for training customer contact staff. It will also serve as a text for students of food science and technology, marketing in its widest sense, retailing, and those concerned with food, its sensory evaluation and its presentation. Although aimed at the food industry, the approach is applicable to all those seeking to understand the environment and expectations of the customer or client.

Understanding the mechanisms of expectations will increase knowledge of ourselves as shoppers and will increase the understanding of the industry as to how to treat customers.

Contents

7. Expectations, Color and Appearance in the Store

8. Expectations, Color and Appearance in the Food and Drink Consumption Environment

9. Expectations, Color and Appearance of Food

1

Expectations, Color and Appearance

EXPECTATIONS

Expectations govern our lives in general and particularly our attitude to food and the food scene. They play a key role in any decision making and subsequent action process. Expectations arise either from belief alone or from a sensory stimulus. An example of the former is the child's belief in Father Christmas or Saint Nicholas that leads to the expectation of receiving presents at Christmas time. Belief expectations form a great part of religious behaviour. Expectations also arise from sensory stimulation. For example, the smell of fresh coffee or freshly baked bread stimulates expectations of the next meal or break in the routine of life. However, by far the greatest stimulus provided by the environment is that which acts on the visual sense. This book is devoted to a consideration of expectations arising from the visually perceived environment which itself consists of colour and appearance of materials in the scene.

We purchase an object not for the sake of the object itself, but for the expectations we have of the object. Color and appearance of the object and of our environment create expectations that affect the way we feel and behave. This is true of all purchasing opportunities and environments, and certainly applies to food and the environments associated with food. Manufacturers attempt to use the visual impact opportunity provided by the advertisement, pack, and label to arouse appropriate feelings within the customer. The color and appearance of the food itself indicates species, variety, quality, and its place within our customary eating pattern. Appearance factors of the store result in customer comfort or discomfort. The meal environment can be deliberately fabricated to inform and reinforce the context of the meal. In food, store, and restaurant the designer uses the scene to communicate with the shopper and eater. Thus appearance both determines and dominates expectations.

Humans, like other animal species, use the visual sense both to detect food and to scan it for quality of eating expectation. Indeed, color vision evolved so that we could detect ripe fruit from its background of green leaves. This set up our color vision to play a vital part in our food selection, appreciation, and environment appraisal processes. Our first images and impressions of a scene are obtained through vision because this acts at a distance. Hence, we form the first judgement of a restaurant when looking through the window. We do not have to listen, smell, touch, or taste before we make our initial, and possibly only, impression of quality and fitness for purpose. Although flavor, or perhaps texture, is normally the final arbiter of quality, food must first pass the appearance test.

Appearance systems evolve and develop, whether the systems concern the evolution and growth of plants and animals, shopping and eating systems, or the systems we ourselves employ when dealing with food supply and consumption. Plants evolved before animals and developed color pigments for converting solar energy into food, for pollination, and for protection (the pigments acting as antioxidants by disposing of free radicals). In turn, animals depend on plant pigments for successful development and protection of their metabolisms. The value of a variety of natural food pigments to human diet is indisputable.

When examining a food material, our visual sense assesses the appearance mechanisms of structure, surface texture, color, translucency, gloss, and patterning. Judgements formed lead to specific expectations and decisions of whether to buy, whether to carry on cooking, or whether to eat. Decisions of what to eat, however, are often governed by what we can buy with the portion of income set aside for food. However, what we eat also defines our personal origins and evolution, our tribal upbringing and development, our nationality, and our position in the social hierarchy. For example, tribes isolated by geography and lack of cheap transport eat what is available locally. Class distinction within the tribe is signalled by the quantity and quality of local or imported foods. In the ancient Roman Empire, for example, the poor drank water, the rich wine. Lack of proper equipment, shortage of fuel, and fire risk in crowded areas limited the availability of cooked food to what could be afforded from cook shops. Most could eat locally reared flesh, vegetables, and fruit, while the richest could dine on imported produce. Extravagance is power and the expression of power and to offer cheap foods to guests is an insult that in turn lowers the esteem and expectations of the donor.

When we open our eyes, immediately there is a scene out in front of us, perhaps a person, an advertisement, a meal, or a store. The lighting defines the stimulus (that is, the extent of the visible scene) that is converted via the retina into our personal images and expectations. Before designers start on a scene creation process they ought to be aware of the expectations users will have of the scene. If they are not then customer feelings of frustration, discomfort, or even helplessness can be aroused. Expectations are formed within our immediate and considered images of the total appearance of the scene. Hence, we make decisions on the basis of these appearance images. Images and expectations are not static and stable; they change with our individual experiences. An unfortunate event associated with a particular space or product will influence our expectations when we meet that or similar space or product again. Similarly, if we have been disappointed with a product for which we had high expectations we will be less optimistic of the performance next time we meet that or a similar product.

Art, science, design, and technology form a continuum of experience linking the scene and its perception. When we view a scene, perhaps looking into a restaurant or scanning the contents of a freezer cabinet, appearance images are formed, some of which attract us favorably, others driving us away. The process from scene to image can be thought of as an information transfer process, in which information from the scene, that is, the scene physics and design working in tandem, is transferred into perceived images.

Images are of two types, immediate or gestalt images and considered images. From our immediate images we recognise a scene, we are aware that we fit (or do not fit) into the scene and also that we may like the scene. An image of "This restaurant looks clean and comfortable" can lead directly to an immediate expectation that "I shall enjoy my meal." The second type of appearance image is more considered and may be sensory, emotional, or intellectual in origin. A sensory image may be "This room smells like a bakery." Such an image may lead to the expectation that "We are going to be given freshly baked bread with our meal." An emotional image of "I hate high ceilings when I am eating an intimate meal" may lead to a lowering of expectation of the meal itself. An intellectual image of "The curtains do not match the carpet" may lead to a lowering of expectation as to the quality of the meal about to be served. Hence, if the environment leads to a positive cognitive set it will lead to increased sales. The designer's job will be easier when we can go some way towards understanding the formation of such images, which in turn will lead to increased customer delight.

Successful design therefore presupposes an understanding not only of scene creation but also of the subjects viewing the scene. The philosophy is common for interior designers, architects, product designers, and designers of packaging and advertisements. We can call these viewer perceptions and expectations the *total appearance* of the scene.

FACTORS CREATING EXPECTATIONS

The *information transfer process* describes the transfer of information from the physics of the illuminated scene through the viewer's physiological processes to the creation of images and expectations, the *total appearance* of the scene (Hutchings, 1999).

First, consider the *viewer-dependent variables*. Images experienced by every customer walking around the store or sitting in the restaurant are influenced by many factors affecting them personally. Sensory signals arising from our receptor mechanisms are processed through our inherited and learned responses to specific objects and events and influenced by our immediate condition and environment.

> *Receptor mechanisms* consist of inherited and acquired sensory properties. When we are in an observing mode our main source of information intake is through our visual sense. This is comprised of our ability to see precisely and accurately the geometry of the scene as well as our ability to see in color. These in turn are affected by perception phenomena, described in Chapter 2, coupled with aging effects and input from other senses.

> *Inherited and learned responses to specific objects and events* are shaped from our culture, memory, preferences, and responses to fashion, as well as physiological effects and psychological responses to, for example, specific elements occurring in the scene, such as color.

> Our *immediate condition and environment* includes geographical factors, such as climate, landscape, and seasonal change, social factors, such as crowding, personal space, and our degree of awareness, and medical factors, such as survival and need, our state of well being, and the protection we require.

Secondly, *scene-dependent variables* consist of the form in which the designer arranges the materials of the scene and the physical properties of those materials.

> The *designer* designs for function, communication (i.e., for identification, safety, or symbolism), and aesthetic reasons, which exist in different numbers of dimensions: one dimensional, i.e., writing; two dimensional, e.g., painting; three dimensional, e.g., architecture; and four dimensional, e.g., performing arts.

> The *material physics of the scene*, or *material properties*, consist of optical properties, including spectral reflectance, transmission, goniophotometric, (that is, properties that change with angles of illumination and viewing), and pattern, physical form, including shape, size, and surface texture, and temporal aspects, including movement, gesture, rhythm, and aging; and the lighting of the scene, which includes illumination type (primary, secondary, or tertiary), spectral and intensity properties, directions and distributions, and color-rendering properties.

Both scene-dependent and viewer-dependent variables affect the images we perceive of the scene in front of us. Perceptions are of two types: *basic perceptions* and *derived perceptions* or *expectations*.

Expectations for people and places are very similar.

Visually assessed safety involves safety of body and safety of mind*
> For places, perhaps "This looks dirty" or "This ladder is very steep" or "The noise in there will drive me mad."
> For people, perhaps "Will this person beat me up?" or "He will bore me to death."
> For food, perhaps "This looks mouldy."

Visual identification.
> For a place, perhaps "Is this a restaurant or a pub?"
> For a person, perhaps "What does this person do for a living?"
> For a food, perhaps "Is this water or is it vodka?"

Visually assessed usefulness.
> For a place, perhaps "Will this store sell my brand?"
> For a person, perhaps "Will this person be able to help me?"
> For a food, perhaps "Is this what I need to eat or drink at this moment?"

Visually assessed pleasantness.
> For a place, perhaps "This is a nice environment."
> For a person, perhaps "How friendly will this person be?"
> For a food, perhaps "How tasty will this plate of food be?"

Visually assessed satisfaction.
> "How satisfied will I be when I have finished my business in this place or with this person."
> For a food, perhaps "I will enjoy this" or "This looks full of cholesterol" or "This meal will not fill me up."

*Note, this is a completely general model and, for completeness, would include safety of spirit. For example, those holding certain religious beliefs will not sit down to eat with those not of the same faith

FIG. 1.1. The five major types of expectation, with examples for places, people, and food.

Basic perceptions comprise the visual structure of the scene, surface texture of elements of visual structure, color, translucency, gloss, and the uniformities of each element, and movement and changes with time of the visual structure. Basic perceptions normally carry the language of visual criticism such as "This is not big enough" or "It is too rough" or "I wanted one in green" or "He moves very slowly."

Expectations, or derived perceptions, are descriptions involving the use of everyday words, mostly in terms of silent and unexpressed questions and observations. These expectations change according to the occasion and are individual only to our needs of that moment. There are five types of expectation that apply to all perceived scenes and events, as illustrated in Fig. 1.1.

Two other specific expectations form part of the visual identification of foods:

Visually assessed flavor, perhaps "Will this taste sweet or savoury?"
Visually assessed texture, perhaps "This looks crunchy."

Similarly, specific identification expectations arising from the sight of a food advertisement or pack are:

Visually assessed flavor, i.e., what flavor(s) are evoked by the advertised food?

Visually assessed texture, i.e., what texture(s) are evoked by the advertised food?

Specific expectations resulting from stranger/stranger interactions are:

Visually assessed recognition of a person 'type' and associations with respect to previous encounters with this 'type' of person.

Visually assessed convenience, e.g., must I engage with and do I have time to engage with this stranger?

For façades, specific identification expectations are:

Visually assessed recognition of façade logo and associations with respect to a known brand, cuisine, or brewer.

Visually assessed convenience, i.e., is it convenient for me to partake of the advertised cuisine or brew at a price I wish to pay?

For a store-specific identification, expectations are:

Visually assessed associations, e.g., do I associate particular irritations or concerns with this brand or type of store or do I associate it with a particular cost structure?

Visually assessed convenience, e.g., does the store look busy, is the carpark full, have I the time to shop?

Specific identification expectations for spaces such as restaurants and pubs are:

Visually assessed associations, e.g., will I obtain a particular food/drink in an environment of suitable cleanliness, comfort, privacy, and quality and price here?

Visually assessed convenience, e.g., is it convenient for me to eat/drink here, how long will it take to be served?

Generally these are everyday feelings and there is nothing technical about the words we use to describe them. The extent to which these expectations are subsequently confirmed or disconfirmed can have a profound effect on acceptance. Cues gained before consumption give rise to expectations of quality.

Discrepancy between expectation and performance is generally minimised or maximised by the consumer. On the one hand, the expectation will be changed to make it more compatible with the performance. On the other, the difference will be increased and the performance reevaluated to be worse than if no prior expectation had been formed. In the extreme case the sight of inappropriately colored food can make us physiologically sick and a food environment that looks dirty can deter us from eating.

Those desperately searching for ways to make us spend money can use three ways to describe the scene's *total appearance* and *expectations* using *appearance profile analysis*. Thus, *expectations* or *derived perceptions* are described in terms of *basic perceptions*, which can be specified in terms of *physics* or *psychophysics assessments* and *measurements*.

This appearance profile analysis approach defines the origin and derivation of the individual's images, attitudes, and expectations of an object or scene in front of us. The technique provides a mechanism by which images, and hence attitudes to images, can be studied.

Using Plutchik's scales (quoted in Sheth *et al.*, 1999) it is possible to determine and quantify emotions resulting from expectations of a person or scene. Subjects rate how they feel at that moment in terms of three sets of adjectives for each emotion, using scales from, for example, "not at all" to "very strongly". These emotions and scales are:

Fear: threatened, frightened, intimidated
Anger: hostile, annoyed, irritated
Joy: happy, cheerful, delighted
Sadness: gloomy, sad, depressed
Acceptance: helped, accepted, trusting
Disgust: disgusted, offended, unpleasant
Anticipation: alert, attentive, curious
Surprise: puzzled, confused, startled.

Appearances and expectations change according to context. For example, a package may be designed to achieve impact from one glance but some aspects of the design of a restaurant interior can be more subtle as there is plenty of time for the décor to work on the client. Expectations will be discussed relevant to visual perception (Chapter 2), scene lighting (Chapter 3), and in terms of people (Chapter 4), packaging and advertisements (Chapter 5), building façades (Chapter 6), the store (Chapter 7), the restaurant (Chapter 8), and the food itself (Chapter 9).

REFERENCES

Deliza, R., and McFie, H. J. H., 1996, The generation of sensory expectation by external cues and its effect on sensory perception and hedonic ratings, *Journal of Sensory Studies* **11:** 103–128.

Hutchings, J. B., 1995, The continuity of colour, design, art and science—Part 1, The philosophy of the Total Appearance Concept and image measurement, Part 2, Application of the total Appearance Concept to image creation, *Color Res. Appl.* **20:** 296–306, 307–312.

Hutchings, J. B., 1999, *Food Color and Appearance*, 2nd edition, Aspen Publishers, Gaithersburg, MD.

Hutchings, J. B., 2000, The Customer's View of Design, Curtin University of Technology, School of Design, Perth, Australia, Occasional Paper no. 2.

Sheth, J. N., Mittal B., Newman B. I., 1999, *Customer Behaviour*, The Dryden Press, Fort Worth

2

Perception of Color and Appearance

EXPECTATIONS

When we wake from a night's sleep, expectations of the day crowd in on us. We are beset by the positive or negative feelings founded on experiences of the days or months past. Onto these expectations are superimposed others arising through the day from external physical and remembered stimuli. These stimuli arise mainly from our visual sense.

We perceive the world in terms of perceptions and expectations that form the brain's best attempt to interpret the scene. Signals from our eyes are manipulated into a number of impressions that form our personal view of the scene. This chapter consists of a brief description of vision and of color and related perceptions.

VISION

Vision has evolved for reasons of safety. Color vision has evolved so that our ancestors could find food as it presented itself or grew against the contrasting color of leaves (Mollon, 1989). Vision also helps us to detect objects and lights in a scene by providing information on color, form, movement, alignment, and position.

The cornea and lens of the eye form an image that falls upon the retina, a light-sensitive area covering a large portion of the inside of the eye. The iris regulates the amount of light falling onto the retina through control of pupil size. The ability to change focus is governed by the circumferential muscle around the lens. Before reaching the retina, light must travel through the vitreous humour and the preretinal structures. The latter contain the nerve cells and their interconnections that transmit signals stimulated by the retinal image to the optic nerve. The vitreous humour which controls the bulge of the cornea, governs accuracy of focus. It also provides pressure to stabilise eye shape, keeps the light path to the retina clear, and prevents retinal detachment.

The retina acts as a transducer between light entering the eye and the processes of light and color perception taking place in the visual cortex of the brain. The light-detecting elements of the retina are the 120 million rods and 7 million cones, so called because of their shape. Approximately 1 million ganglion cells carry information from the retina to the optic nerve. At the fovea, the convergence is 1 : 1, while at the periphery it is several 100 to one. This degree of convergence determines spatial resolution and sensitivity. Hence, at the faveola, a spot 2 mm in diameter in the center of the fovea containing only cones, there is low light sensitivity but maximum resolution to 1 minute of arc. The number of rods increases to a maximum approximately $20°$ from the fovea. Here, there is high light sensitivity but low spatial resolution. This is the angle used for improved sight in the dark when we use averted vision. Outside this area, color discrimination is zero and the remainder of the retina is probably used solely for the detection of movement.

There are three types of cone, each having a characteristic pattern of response to the wavelength of incoming radiation. Cones are maximally sensitive to reddish (R), greenish (G), or bluish (B) light. Rods function under low levels of illumination, conditions in which cones do not function. Blues then tend to appear brighter and reds darker relative to one another.

Almost half the brain is devoted to visual perception. Information from the retina is processed in two broad systems; one is concerned with identification of the object, the other with relative spatial position. Within the brain's visual cortex there are six specific areas, V1 to V6, responsible for the perception of different aspects of appearance. Area V1 responds to orientation, real and imaginary boundaries, and has some wavelength response. It also detects overlapping features in the scene. Cells in V2 are sensitive to color, motion, orientation, and stereoscopic features. V3 is sensitive to form and depth. V4 is sensitive to color and is the site responsible for the maintenance of color constancy. V5 analyses motion and V6 is responsible for analysing the absolute position of an object in space.

Visual recognition occurs when a retinal image of an object matches a representation stored in the memory. The ability to recognise objects depends on visual neurons being able to form networks that can transform a pattern of points of different luminous intensity into a three-dimensional representation. This enables the object to be recognised from any viewing angle. Neuron cells are responsible for specific recognition tasks and adjacent cells usually respond to very similar feature configurations. These simple shapes form a visual alphabet from which a representation of more complex shapes can be constructed. Some neurons, called face cells, may form a neural substrate for face processing, responding only to faces, whether in life, as plastic models, or on the screen.

Moving objects can be tracked using pursuit eye movement. Considerable neuron interconnections are required to follow objects continuously displaced from one point to another. Two other types of eye movement occur. Continual small movements are needed to destabilise the image and prevent the retina adapting to a continuous stimulus, and larger short movements permit the eye to scan the visual environment.

Many visitors to stores and restaurants are visually impaired to the extent that they have difficulty in finding their way around. It is colour contrast that holds the key to the successful discovery of room size and to wall, ceiling and floor relationships. This also holds the key to successful negotiation of furniture, doors, stairs, switches and toilet facilities. The full benefit of colour differentiation is found when surfaces are given a matt or mid-sheen finish. Glossy and busily patterned surfaces increase confusion (Barker and Bright, 1997).

A problem occurring in supermarkets in particular concerns visual acuity. We are expected to be able to read prices and product data written in small type on colored labels that might be a meter away on the bottom shelf. A study using Japanese 75-year-olds found that few color combinations achieved as efficient a readability than black on white. Of those wearing spectacles 35% could read 2-mm typeface. 50% could read 3-mm, and 90% could read 3.7-m, typeface. However, this was only when subjects were permitted to adjust their own reading distance (Ito et al., 1997).

Color Vision Deficiency

Approximately one man in 13 and one woman in 250 perceive colors in a markedly different way from the remainder of the population. This can have unfortunate results, such as the man who had difficulty picking red cherries because they were the same color as the leaves on the tree. Evidence for color deficiency is not always apparent, for example, in the kitchen. Individuals with poor color perception learn the socially approved color names for many objects where shape and size provide firm clues to identity. Problems may occur with formless food such as sauces though these can always be tasted if there is doubt. The other 92% of the population do not perceive colors in exactly the same way. Eyes, like foods, belong to biological populations in which large variations occur even among 'normal' observers.

Color-deficient vision can be inherited or acquired through damage to the retina or optic nerve. It can arise from illness or ingestion of neurotoxic agents, although the most common acquired form is from cataract and glaucoma in old age. As we grow older our color perception and color sensitivity declines. Ability to discriminate yellows decreases as yellowing in the eye lens increases between the ages of 60 and 90 years. Discrimination in this age range is worse in the blue-green region than the red-yellow and such changes can be extensive. Colored surfaces are perceived by the aged to have less chromatic content than those perceived by younger adults.

Defects arise when responses from the cones, or occasionally rods, are different in some way. The most common causes relate to a deficiency or absence in the G and R cones. The presence of these conditions can be detected by using Ishihara charts, which are based on color confusion. They consist of a series of plates made up of spots of different color. The spots may wrongly be described as being the same color and hence the charts are known as pseudo-isochromatic tests. These can be used to detect the presence of a color vision deficiency, but not necessarily give information as to the type.

Color deficiency is only a significant handicap in certain technical situations. Individuals with one of these conditions will not be able to perform exacting color tasks, such as mixing colors to produce a generally acceptable color match. These skills may be required, for example, when using color additives to achieve a particular color. Other problems have occurred; for example, tomatoes have been picked too early while they were still unripe, and green sweets have been produced instead of red. However, within the food industry, few examples have been reported.

COLOR PERCEPTION

There is a highly complex series of connections between the detector and the fibres making up the optic nerve taking the resulting signals to the visual cortex. Hunt (1998) has outlined a simplified framework. In essence, color information from the retina is obtained from a series of R, G, and B signal differences.

$$R - G = C_1$$
$$G - B = C_2$$
$$B - R = C_3$$

These are transmitted as two signals, C_1 and $(C_2 - C_3)$.

Under full daylight illumination there is also an achromatic signal A, which is a combination of R, G, and B signals, plus the signal from the rods S. The three cone responses are weighted to compensate for the differing numbers of each type of cone $(40:20:1)$ in the retina. That is, the total achromatic signal can be described as:

$$A = 2R + G + B/20 + S$$

In terms of visual sensations of color:

Brightness is the attribute of a color according to which it appears to exhibit more or less light (adjectives: bright–dim).

Hue is the attribute of a color according to which it appears to be similar to one, or proportions of two of the perceived colors red, yellow, green, and blue.

Colorfulness is the attribute of a color according to which it appears to exhibit more or less of its hue.

The value of A determines the *brightness*, yielding a sensation ranging from black through gray to white. For the *hue*, a positive signal of C_1 gives rise to reddish sensations and when it is negative the sensations are greenish. When $(C_2 - C_3)$ is positive, we perceive yellowish sensations and when negative they are bluish. The particular *hue* may be indicated by the ratio of C_1 to $(C_2 - C_3)$, and the *colorfulness* by the strengths of C_1 and $(C_2 - C_3)$.

In the bright light of the kitchen or outdoors a ripe red tomato on a white plate has a strong red *hue* of high *colorfulness*. When it is taken into the lower light levels of the restaurant it will have a lower *colorfulness* because its *hue* will be less strong. Also, the white plate will have a lower *brightness*. The lower *colorfulness* of the tomato in the reduced light level is seen to be caused by the lower level of illumination, characterised by the lower *brightness* of the white. Hence the *chroma*, that is, the *colorfulness* relative to the *brightness* of the white, is unchanged. Thus, the *chroma* is related to C_1/An and $(C_2 - C_3)/An$ signals (suffix n refers to responses for the white).

Lightness is used to describe the *brightness* of objects relative to that of a similarly illuminated white. Whereas *brightness* depends on the absolute magnitude of the achromatic signal A, the *lightness* is seen to be related to A/An, where An is the *brightness* of an appropriately chosen white.

Light falling onto the tomato will vary greatly over its surface depending on angles of incidence. Only in a few places will it be possible to judge the *brightness* of a similarly illuminated white, and hence only in these places will it be possible to judge *lightness* and *chroma*. However, it is possible to judge *colorfulness* relative to the *brightness* of the same area. This defines the attribute of *saturation*. The uniformity of the color across the whole surface of the tomato can be judged by *hue* and *saturation*. Saturation may be related to the magnitudes of the C_1/A and $(C_2 - C_3)/A$ signals.

These three relative subjective terms are defined as:

Lightness, the *brightness* of an area judged relative to the *brightness* of a similarly illuminated area that appears to be white or highly transmitting (adjectives: light and dark).

Chroma, the *colorfulness* of an area judged in proportion to the *brightness* of a similarly illuminated area that appears to be white or highly transmitting (adjectives: strong and weak).

Saturation, the *colorfulness* of an area judged in proportion to its *brightness*.

The six subjectively judged attributes may be used to describe our perception of our environment whether it is in the home, the supermarket, or the kitchen. Hunt's approach may be applied to the assessment of complex scenes.

Color Adaptation and Color Constancy

Lightness and color constancy contribute to a more stable perceived world. Surfaces will appear equally light, relative to the surroundings, over the very wide range of illumination we encounter from the dark of night to the full sun of daytime. This *light adaptation* allows *light constancy* to be possible.

Color constancy also occurs. Whatever the light, from dawn to dusk, coal still appears black and leaves green. When a green object is brought indoors from bluish natural daylight to an area of reddish tungsten lighting, it reflects yellow–green light. However, this change does not appear to apply to the object. The color change is recognised as belonging to the light and we perceive the object color relative to this. That is, the object appears almost as green as it did in daylight. As we become adapted to the new illumination, light reflected from the object is perceived to have almost the same color as when daylight was reflected from it. This phenomenon is *object–color constancy*. However, color constancy is not absolute. Viewing fresh beef under reddish lighting is sufficient to conceal the first visual signs of the brown metmyoglobin, a characteristic of older meat, which would be revealed under bluer lighting.

White tends to appear white whatever the light shining on it as long as our eyes are adapted. In a scene containing a series of achromatic colors (white or gray) the brightest will be seen as white, while the others will appear to be gray. If this white is removed, the next brightest gray will come to be called white. That is, our perceptual system notes the highest brightness achromatic in the field of view as white and then relates all other colors to it. In the absence of grays, other colors may act as reference whites also. This phenomenon occurs for a wide range of lighting, certainly all common daylight-type fluorescent and tungsten lamps. As the perceived white is a reference for all other colors, these tend to be perceived as constant under different illuminations. Grayness is perceived as a relative reduction in the proportion of light reflected; that is, there must be a reference to the total amount of light present. Mistakes in perception can occur if there is no clue to the total amount of light present. For example, if the lightest surface in view in a room is gray it may be mistaken for white, and the level of illumination judged lower than it actually is. When someone walks in front of the wall, either their complexion will appear unnaturally vivid, or the wall will suddenly change color from white to gray.

Adaptation taking place on a change of illumination is not complete. The color shift on adaptation is a mixture of an objective shift (caused by a different balance of wavelengths being reflected from different light sources), and a subjective (adaptive) color shift. This adaptive color shift is caused by *chromatic adaptation*.

There is another type of constancy. When we see shadows cast onto a plain plate, we do not immediately think that different colored paints have been used in the finish. We subconsciously assume that it is uniformly colored; that is, we make an immediate adjustment to allow for the change in illumination falling on the different areas. Similarly, parts of the surface of a green apple in a bowl of fruit may be in shadow, or it may be altered by reflections from adjacent differently colored fruit. However, we perceive the apple to be uniformly green. A bruise on the apple will probably be seen because we can recognise that the damaged surface is not the same color or shape that would result from a shadow or reflection from other items in the bowl. When we look at our red car in the drive we do not examine the color in detail. If other factors, such as size and shape and design, are appropriate it must be the same color as it was before. That is, for this type of constancy our memory and other appearance properties govern the color perception. This might be termed *total appearance constancy.*

Problems caused by chromatic adaptation can occur when serving food. Change of illumination between kitchen and dining room may result in the customer perceiving different colors from those intended by the chef. Similar effects occur with restaurant decor, particularly with dimly lit decor. If this possesses the wrong balance of brightness or color, the derived perception or expectation of the decor is that it may be dirty and therefore unhygienic.

In survival terms, light and chromatic adaptation have significance for humans in maintaining the stability of some perceptions in circumstances of an ever-changing external environment.

Contrast Phenomena

Lightness contrast is well known; it is the reason we can see stars at night, but not in daytime. Contrast between the lightness of star and sky is greater on a dark night than in daytime.

The color of an object in a complex scene depends, through various contrast mechanisms, on other colors occurring in the scene. When the eye is exposed to one color, say red, the red cones become saturated. When a second color is substituted, the signal from the red cones will tend to be low in comparison to that from the green cones. The resulting $(R - G) = C_1$ signal will be more negative and the object will appear too green. So, a yellow area can appear green if the eyes are first adapted to red for a short while. This is *successive contrast* and requires no fixation for it to occur. *Successive contrast* also occurs in *after-image* effects when, say, our eyes have become fixed on an orange tablecloth. When food is subsequently placed in front of us it will be tinged with the complementary color, blue. Such complementary color effects can be uncomfortable when, for example, sorting green peas on a conveyor belt induces perception of red spots.

The color appearance of a color changes when another color is placed alongside. *Simultaneous contrast* occurs when we look at colors placed side by side. One color assumes some of the hue of the complementary of the adjacent color. For example, a gray circle on a green background will appear pinkish. When carrot and potato occur together on a plate, the potato has conferred upon it the blue that is complementary to the carrot's orange. When complementary colors are placed together both are exaggerated. The orange of our carrot is intensified when served on a blue plate. In the window of the butcher's shop, the presence of green herbs around a piece of raw beef enhances the redness of the meat. Effects of adjacent complementaries can be uncomfortable in their boldness.

Other types of contrast are also of interest to designers. *Border contrast* occurs when changes in the border zone between two colors tends to increase the perceived difference between the colors. *Contour contrast* is effective in enhancing apparent brightness or volume. For example, if, within a uniformly colored surface, a small area of shading marks the circumference of a circle, the surrounded circle will appear to be a lighter color. *Vibration*, an intense shimmering effect, can be induced along contours, particularly when complementary hues of approximately the same lightness are involved. In *optical mixing*, Pointillist artists exploit the fact that, from a distance, the eye cannot perceive the edges of small shapes. These tend to blend into each other, the result being seen as a mixture. An illusion of *transparency* occurs when there are three adjacent colors if the middle one is a mixture of the other two. Other aspects of contrast are discussed in Chapter 5.

These and other complex visual effects appear to be caused through the mechanisms of simultaneous and successive contrast, and an interaction of excitatory and inhibitory mechanisms in adjacent cones. Contrast phenomena arise as a result of the different functions within the visual cortex of color selectivity, contrast selectivity, temporal properties, and spatial resolution. These result in the different visual functions of color, depth, movement, form, and orientation perception.

Color Memory

Color and appearance memory are used when making decisions about food quality. We judge freshness, ripeness, cleanness, naturalness, extent of cooking, bruising, and spoilage. Of the environment we judge perhaps cleanness and safety. These judgements are governed by our upbringing, culture, and how we are feeling at the time, as described in Chapter 1. Food color and appearance are compared with mental images formed from past experience. If something about a food or environment does not correspond with our memory of wholesomeness and hygiene, the product may be left on the shelf or the plate.

The presence of a mental image is a significant factor in how quickly we recognise an object when it appears in front of us. If the real object is, say, differently aligned from the image, our recognition speed is decreased. Similar mechanisms exist by which we are disappointed if the quality of the object does not match the quality pictured by our mental image, our expectation. Once an image is formed it can in some ways function like the object itself and activate other types of neural mechanism. Images, once formed, may come to modify the perception. This has profound cross-cultural implications.

Humans do not memorise colors well. Some can memorise with accuracy for up to 30 seconds and perhaps with small errors for some hours. However, a tie purchased to match a shirt in the wardrobe will probably not be a good match on the return home. When estimating colors from memory, we tend to overestimate the dominant chromatic attributes of certain objects. Also, in most cases, saturation and lightness increase in memory. For example, grass is greener than it really is, skies are bluer, and bricks redder. Makers of film build these preferences into products; hence, for example, pictures in the fast food restaurant can be more chromatic than the food itself.

Increasing use of photographs and video images to represent food calls for consideration of *quality* and *naturalness* as attributes of hard copy or on-screen pictures. Naturalness is the degree of similarity of the picture to the memory prototypes of the object itself. Naturalness judgements of colors in a whole scene are determined by the naturalness of the most critical object in the scene, such as skin tone. As already noted, observers normally prefer scene elements of a picture to be more chromatic and colorful than the original. This holds particularly for low chroma scenes. There is a strong link between reproduction image quality and naturalness, especially in images of natural scenes (Fedorovskaya *et al.*, 1997).

Color memory is used for tasks vital to our well-being. From an evolutionary point of view, we must memorise the combination of food plus its color and develop the ability to recognise particular stages of natural ripening and spoilage processes. Even this ability may not be tested very often as in most cases contrast colors are present during our visual ripeness and spoilage judgements. Examples include bruising of fruit or the appearance of a slight green tinge to meat. Hence it is likely that it is the aptitude of the visual system in judging color contrast that accounts for many of our food quality decisions. In the development laboratory, judgements of product quality are frequently made from a visual image. Therefore, when making critical or disciplined judgements of flavor and texture, it is essential to eliminate the presence of visually arising cues.

Color Harmony

Color harmony occurs when two or more colors seen in neighboring areas produce a pleasing effect. It is a purely personal reaction to a perception. Harmony is used and exploited by designers in their bid to communicate and create aesthetically pleasing scenes. Appreciation of the designer's efforts depends on the viewer's learning, appreciation of fashion, whether we have become tired of, or have begun to appreciate, certain color combinations, their shapes, the number and saturation of colors, layout, and relative sizes of colored areas. Harmony is also affected by the absolute angular size of areas. For example, a Roman mosaic multiplied ten times would produce a garish effect. We also back away from large paintings to obtain a better view. Undesirable contrasts can occur through the chromatic adaptation and area effects discussed above under adaptation and constancy. Manuals containing examples of suggested color harmony for use in various design situations have been published (e.g., Kobayashi, 1998).

Color harmony is personal. However, a set of principles has been collected, which, although not scientifically verified, might act as a guide to the selection of pleasing color combinations (Judd and Wyszecki, 1975). There are four major principles:

Color harmony results from the juxtaposition of colors selected according to an orderly plan that can be recognized and emotionally appreciated.

The principle of familiarity is relevant; we like what we are used to. The harmonies produced by nature provide examples.

Any group of colors will be harmonious if, and to the degree that, the colors have a common aspect or quality. That is, the colors are more or less alike. Differences will be reduced if it is seen that they have much in common. The common aspect might be a series of colors of constant hue, constant lightness, or constant darkness.

Unambiguity in the selection of colors is appreciated.

Harmony or color schemes can be classified (Radhakrishnan, 1980).

Analogous color harmony arises when two or more colors, which are similar or related to each other in the spectrum, occur together. This basic scheme may be used with variations of hue, brightness, and saturation.

Complementary color harmony occurs when two or more colors, which are dissimilar to the extent of being complementaries, exist side by side or near together.

Monochromatic harmony is the simple harmony of one chromatic color. Every hue may be graduated into a series of shades by the addition of black, gray, or white.

Achromatic harmony is based on whites, grays, and blacks.

Undesirable contrasts can occur through the chromatic adaptation and area effects discussed above under constancy. Some aspects of the application of color contrast and harmony to packaging are discussed in Chapter 5.

Color harmony has had attributed to it many different meanings, systems and sets of rules (Burchett, 1991). However, probably the most consistent rule for the designer is not to be enslaved by rules.

COLOR-RELATED PERCEPTIONS

Color, translucency, and gloss are important visually perceived properties of all scenes including foods. We employ our color vision in the detection of these (as well as other) material properties. When light hits the surface of an object it may be reflected from the surface as well as transmitted into or through the material.

Light is reflected from the surface both regularly and diffusely. Light regularly reflected leads to perceptions of *gloss* and gives visual clues to the *color of the light source*. Light diffusely reflected contributes to perceptions of the *color of the material*.

Light transmitted into the material is scattered and absorbed within the body of the material. Visually significant scattering within the material leads to perceptions of *translucency* of the material and contributes to the perception of its *color*. Materials that permit no transmission into the surface are *opaque*.

Light may be transmitted through the material in a regular or diffuse manner. Regular transmission leads to perceptions of *transparency*. Transmission that is diffuse leads to perceptions of *translucency*.

All these phenomena contribute to perception of the *color* of the material.

All foods have color. Although some foods can be regarded as opaque, most are transluscent. All solid and semi-solid foods are glossy to some degree. Liquid foods may be visually transparent or translucent. Interaction of color, transparency, translucency, opaqueness, and gloss occurring in food and food display are discussed in Chapter 9.

UNDERSTANDING EXPECTATIONS IN TERMS OF SPECIFICATION AND VISUALISATION

Using a food as an example, we can describe the eater's expectations in terms of psychophysics or visual criticism and physical measurements. Expectations can be summarised under the headings of visually assessed safety, visual identification, visually assessed flavor and texture, visual satisfaction, and the change of these visual properties with time. The psychophysics or visual criticism is described in words, but most can defined instrumentally. The psychophysics attribute classes are visual structure and substructure (including sizes and shapes), and, for each element of the visual structure, surface texture, color, gloss, translucency, and patterning. For example, visual criticisms that this meat is raw or cooked or burned, that is, black, can be measured in terms of color. The color measurement can be in terms of a reference to a color atlas and, because we have instruments that relate to verbal descriptions of color itself, that is, whether it is red, green, or blue, it can be specified instrumentally. Similarly, we are in the habit of verbally describing qualities of gloss, translucency, and patterning. It will soon be possible to apply modern digital measurement technology to the specification of all these psychophysical factors (Hutchings, 2002).

This description has been for food, but the same principles apply to the measurement understanding of expectations of people and spaces.

Connotative Meaning of Objects

It is possible to determine the connotative meanings of objects using the semantic differential technique (Osgood *et al.*, 1957). Relationships existing between objective properties of materials and scenes and their psychological meaning can in this way be determined. There have been applications to architecture, but there appears to be no reason why its use cannot be extended to the food product, package, or people. This principle has been applied to eating and drinking venue interiors in Chapter 8. As an example, one application involves subjects making judgements of three independent dimensions, each comprising a number of polar adjective pairs. The dimensions with their relevant polar pairs are:

> *Evaluation dimension*: beautiful–ugly, pleasant–unpleasant, harmonious–dissonant, meaningful–meangingless, cheerful–sad, and refined–vulgar.
> *Activity dimension*: energetic–inert, tense–relaxed, dynamic–static, interesting–boring, warm–cold, and fast–slow.
> *Potency dimension*: rugged–delicate, hard–soft, tenacious–yielding, strong–weak, tough–tender, and masculine–feminine.

The connotative meaning of an object or concept is obtained by asking respondents to indicate, say, on a seven-point scale between each polar adjective pair, the point that best describes the relationship or association between the concept and the adjective. Factor analysis is then used to derive ratings for the three dimensions. These dimensions can be plotted orthogonally in Cartesian coordinates, thus developing a semantic space. This technique can be used to compare the meanings of two concepts, say a design concept and a visual surface property. The formula used for this is:

$$D_{ds} = [(E_d - E_s)^2 + (P_d - P_s)^2 + (A_d - A_s)^2]^{0.5}$$

where D_{ds} is the distance between the meaning of the design concept and a visual property of a surface in semantic spatial units, E is the evaluative coordinate, P is the potency coordinate, A is the activity coordinate, d is the subscript indicative of the design concept, and s is the subscript indicative of the visual property of the surface.

The technique can, for example, be a rational basis for the selection of a particular surface from a number of alternatives. The surface selected is the one that matches the design concept in connotative meaning, that is, one for which D_{ds} is minimized.

In an architectural example the connotative meanings for unglazed brick textures, colors, and sizes have been determined. Judgements of texture showed that, in general, brick textures are neither good nor bad, but they may be strong or weak, or active or passive. Few brick colors were either good or bad. Moderately orange bricks were determined to be the most active, brownish-pink the most passive. Grayish-reddish-brown bricks were the strongest, pale yellowing-pink the weakest. The potency ratings given to size varied considerably more than the other two dimensions. The intra-observer reliability, and the agreement between two groups of cooperative and well-motivated observers were found to be high (Burnham and Grimm, 1973).

Methods have been suggested for pre-testing such market stimuli as product benefit combinations, package designs, brand–price combinations, advertisements, and special offers. Marketing problems can be broken down into specific areas, for example, measuring the relative importance of a group of product benefits, or selecting a package design that best relates to the psychological imagery of selected benefits, or determining what point-of-purchase display materials, brand name, and pricing strategy to employ in the market introduction. All these problems involve non-numerical, judgemental responses to multi-attribute marketing stimuli.

This type of problem may be approached using conjoint measurement. This involves using conventional market research techniques to obtain a list of product benefits thought to be the most important to the market. Conjoint measurement algorithms require respondents to rank order fractional factorial designs that incorporate the product benefits. Benefit interactions may also be included in the analysis (Green, 1973).

Color Order Systems and Atlases

Undoubtedly, color plays a major part in judgements of quality. Although the eye is good at discriminating between colors, our capacity for remembering them is comparatively poor. Hence, it is useful to have a system of reference whereby colors can be specified and one method is to match the sample to a color chip obtained from the atlas of a color order system. A color atlas consists of a book containing pieces of colored card; the color order system is the method by which the colors in the atlas are laid out and described.

Advantages of using an atlas include:

Colors are visualised.
Comparatively low cost.
Stability and consistency; although chips may easily become contaminated commercial color atlas chips are reproduced to precise and accurate tolerances.
Portability.
Ease of communication, say, between designer and producer.

Disadvantages include:

Necessity for a standard viewing and lighting system.
Necessity to check that the observer has normal color vision and is consistent at the task of matching colors.
The comparatively large steps between adjacent chips means that often only an approximate match is possible, but interpolation is possible in established color order systems.
Saturated colors may be missing from the atlas because they are difficult to reproduce in paper media.
Difficulty of matching intense, dark colors.

Many color atlases have been produced, but two with their color order systems will be described. The most commonly used atlases are the Natural Color System (NCS) and the Munsell System. Atlases are used for direct comparison with the sample, but the scales of each can also be used independently of the atlas by skilled observers.

The NCS atlas uses Hering's postulate that all colors may be placed in a system with reference to six elementary color sensations. These are the achromatic sensations of whiteness (V), blackness (S), and chromatic sensations of yellowness (Y), redness (R), blueness (B), and greenness (G). These are elementary hues where, for example, unique yellow is defined as being neither reddish nor greenish. No color is normally seen as yellow and blue at the same time; hence, this is an opponent air. The same reasoning applies to the unique red–green pair. Any one color may be specified in terms of the percentages of two chromatic and two achromatic attributes.

Thus, a particular color may have the NCS description 2030-Y90R. The *nuance* of this color is 2030, that is, the degree of resemblance with *black* (S) and the maximum *chromaticness* (C). This color contains 20% black and 30% chromaticness. The *whiteness* of the color need not be shown separately as it can be calculated from the equation:

$$W\% = 100\% - S\% - C\%$$

The *hue* notation describes the degree of resemblance of the color to the two chromatic elementary colors, in this case Y and R. Y90R indicates a yellow color with 90% reddishness and 10% (i.e., 100% − 90%) yellowishness. Pure gray colors lack hue and are specified in terms of nuance followed by the letter N (i.e., neutral).

Each page of the NCS atlas contains physical samples of a single hue and a triangle having a vertical neutral (black to white colors) axis and colors of different nuance and chromaticness. Colors of higher chromaticness are further away from the neutral axis. The exact color can be depicted on an NCS three-dimensional color solid consisting of a hue circle and all hue pages.

Similarly, the Munsell atlas consists of a hue circle and pages of constant *hue* (H). The *chroma* is zero at the achromatic axis and increases in visually equal steps to /10, /12, /14, or greater for particularly saturated colors. Chips are arranged so the vertical axis of the page represents an increase in value (V) from black = 0 to white = 10, the horizontal axis an increase in C that is an increase in hue content and departure from gray. Each page is a vertical section through the model, taking one hue at a time. The Munsell description of a yellow-red color of hue 3YR, value 5/, and chroma /6 is 3YR 5/6. Interpolation between whole units is possible, and single decimal places may be used where appropriate.

Hence, colors can be specified by color matching to a particular color chip in the atlas or colors can be visually scaled with reference to a particular set of color dimensions, for example, those of NCS or Munsell. The former is more straightforward to learn.

Color and color appearance can also be measured instrumentally using a reflectance spectrophotometer or a tristimulus colorimeter.

Color Image Scale

We are undoubtedly sensitive to color, but just how we are sensitive and what meanings are conveyed by color is often a matter of confused speculation.

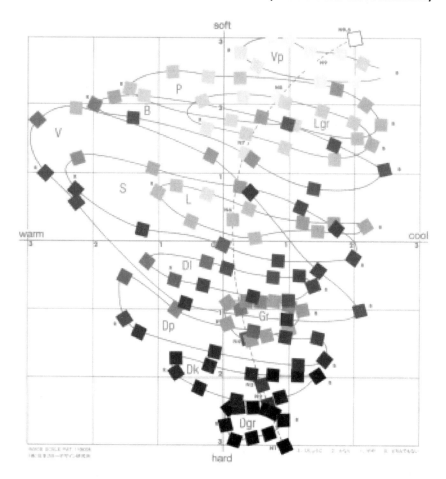

FIG. 2.1. The Single Color Image Scale developed by Shigenobu Kobayashi.

However, Shigenobu Kobayashi (1998) has developed a color image scale that organises colors and color combinations according to their images in everyday life. This was developed for Japanese clients of Japanese designers. Undoubtedly some of the parts of the scale are as applicable to Westerners as they are to the Japanese but it has not been totally confirmed for nonoriental areas. Nevertheless it forms a useful guide to relationships between colors scaled in common terms as well as their group relationships with defined emotions. Examples of these emotion groups are 'pretty,' 'elegant,' 'classic,' and 'romantic.'

The Single Color Image Scale is shown in Fig. 2.1. This depicts colors in terms of two scales, *hard* to *soft*, and *warm* to *cool*. Kobayashi further describes the colors, referring to letters on the figure, as:

Figure 2.1

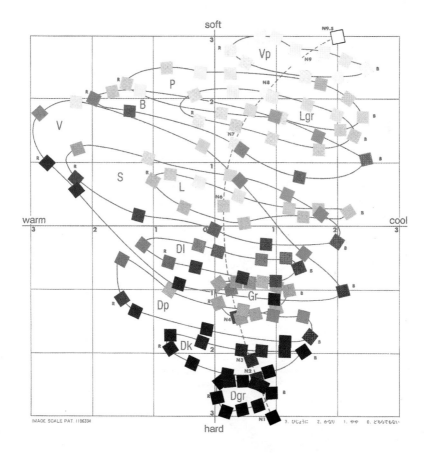

Figure 2.2

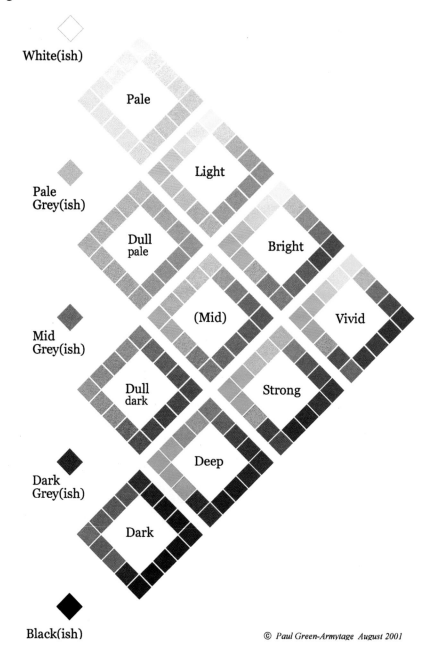

White(ish)

Pale

Light

Pale
Grey(ish)

Dull
pale

Bright

Mid
Grey(ish)

(Mid)

Vivid

Dull
dark

Strong

Dark
Grey(ish)

Deep

Dark

Black(ish)

© *Paul Green-Armytage August 2001*

Vivid tones are V vivid or S strong.
Bright tones are B bright, P pale, or Vp very pale.
Subdued tones are Lgr light grayish, L light, Gr grayish, or Dl dull.
Dark tones are Dp deep, Dk dark, or Dgr dark grayish.

Vivid tones (V) are bold, clear, full of life; they attract attention and are sharp and lively. Strong tones (S) are slightly duller than V tones; they are sturdy and substantial and convey a practical feeling.

Bright tones (B) are bright and clear, like sparkling jewels reflecting light; they convey images of a sweet flavor. Pale tones (P) create a dreamlike atmosphere; they are pretty, sweet, and dreamy. Very pale tones (Vp) are light colors that convey feelings of softness and delicacy.

Light grayish tones (Lgr) are simple, gentle, quiet colors. Light tones (L) convey a mild and charming image. Gray tones (Gr) have little pigmentation; they are simple, quiet, and elegant. Dull tones (Dl) are quiet, sophisticated, and old-fashioned.

Deep tones (Dp) are full-bodied, tasteful, and convey a substantial feeling. Dark tones (Dk) contain a subtle tinge of hue; they are resonant colors and convey an atmosphere of stability. Dark gray tones (Dgr) are close to black, severe, serious, austere, and convey a feeling of precision.

All achromatic colors have a cool image. White is refreshing while black is heavy and severe. The grays range from light to dark and can be used to convey a calm feeling. They provide effective foils to chromatic colors.

Color combinations can also be described in terms of the color image scale. An example is shown applied to food packs in Fig. 5.2.

Colour Zone Diagram

The color solid of the Natural Color System (NCS) provides the structural skeleton for the Colour Zone Diagram developed by Paul Green-Armytage (Green-Armytage, 2001). The three-dimensional nature of NCS is presented in two projections—a plan view (a color circle showing a sequence of hues) and part cross-section (a color triangle, which shows a set of nuances). The Colour Zone Diagram, shown in Fig. 2.2 is part hue circle (each circle actually pictured as a square) plotted onto a nuance diagram. Thus the color solid is shown as a two-dimensional array of zones that can be clearly understood and visualised.

There is a vertical axis from *blackish* through *grayish* to *whitish*. Each hue square is labelled according to the common color feature of the square. For example, as the high chroma colors, that is, the *vivid* colors, are successively diluted with white they move through the sequence *vivid-bright-light-pale-white*. The zones change character similarly when diluted with black, moving through

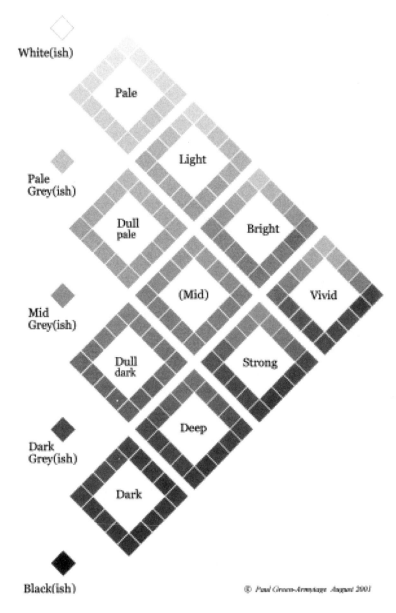

FIG. 2.2. The Color Zone Diagram developed by Paul Green-Armytage.

vivid-strong-deep-dark-black. Zones diluted with shades of gray are labelled *dull pale, dull dark,* and *mid.*

The diagram is not a precise depiction of color space, but it is extremely useful for specifying the color character or zone of a particular, say, architecture or design system under study. For example, it can be used to indicate the different color characters of restaurant facades found in different geographical areas of the world (see Chapter 6).

REFERENCES

Barker. P., Bright. K., 1997, *A Design Guide for the use of colour and contrast,* ICI, Slough
Bloomer, C. M., 1976, *Principles of Visual Perception,* van Nostrund, New York.
Burchett, K. E., 1991, Color harmony attributes, *Color Res. Appl.* **16:**275–278.
Burnham, C. A., and Grimm, C. T., 1973, Connotative meanings of visual properties of surfaces, in: *Sensory Evaluation of Appearance and Materials,* R. S. Hunter and P. N. Martin, Eds., ASTM, Philadelphia.
Davidoff, J., 1991, *Cognition Through Color,* MIT Press, Cambridge, MA.
Fedorovskaya E. A., Ridder Huib de, B. F. J. J., 1997, Chroma variations and perceived quality of color images of natural scenes, *Color Res. Appl.,* **22:**96–110.
Finke, R. A., 1986, Mental imagery and the visual system, *Sci. Am.* **254 (March):**76–83.
Fletcher, R., and Voke, J., 1985, *Defective Color Vision,* Hilger, Bristol.
Green, P. E., 1973, Measurement of judgmental responses to multi-attribute marketing stimuli, in: *Sensory Evaluation of Appearance of Materials,* R. S. Hunter and P. N. Martin, Eds., pp. 139–153, ASTM, Philadelphia.
Green-Armytage, P., 2001, Colour zones, explanatory diagram, colour names, and modifying adjectives, *Proceedings of the International Colour Congress,* Rochester, R. Chang and A. Rodrigues, Eds., 2002, 976–979, SPIE—The International Society for Optical Engineering, Washington.
Hunt, R. W. G., 1998, *Measuring Colour,* 3rd edn., Fountain Press, Kingston-upon-Thames.
Hutchings, J. B., 1999, *Food Color and Appearance,* Aspen Publishers, Gaithersburg, MD.
Hutchings, J. B., Luo, M. R., Ji, W., 2002, Calibrated colour imaging analysis of food, in: *Colour in Food: Improving Quality,* D. MacDougall Ed., Woodhead Publishing, Cambridge.
Ishihara Charts, Kanehara Shuppan, Tokyo.
Ito, N., Yoshioka T., and Imai Y., 1997, Japanese senior citizens' legibility of letter with color, in *Proceedings of the 8th Congress of the International Color Association,* Kyoto, The Color Science Association of Japan, 347–350.
Judd, D. B., and Wyszecki, G., 1975, *Color in Business, Science and Industry,* 3rd edn., Wiley, New York.
Kobayashi, S., 1998, *Colorist, a Practical Handbook for Personal and Professional Use,* translated by Keiichi Ogata and Leza Lowitz, Kodansha International, Tokyo.
Mollon, J. D., 1989, "Tho she kneel'd in that place where they grew...": the uses and origins of primate color vision, *J. Exp. Biol.* **146:**21–39.
Munsell Color Co, 2441 N Calvert Street, Baltimore, MD 21218.
NCS, Scandinavian Colour Institute AB, PO Box 49022, S-100 Stockholm, Sweden.
Osgood, C. E., Succi, G. J., and Tannenbau, P. H., 1957, *The Measurement of Meaning,* University of Illinois Press, Urbana, IL.

Radhakrishnan, K. A., 1980, Colour and textile design, in: *Color 80 Seminar Proceedings, a Colorage Supplement*, 43–48.

Tovée, M. J., 1996, *An Introduction to the Visual System*, Cambridge University Press, Cambridge.

3

Lighting and Illumination

EXPECTATIONS

Light is that part of the electromagnetic spectrum perceived by the eye. We see light in the form of a direct source such as a lamp, but otherwise we can only see it after it has been scattered or reflected by a material with which it has come into contact. The presence of lighting defines the scene in front of us and we use vision to warn and inform us about the scene. If there is no illumination we need other senses such as touch for scene definition. The most common light emitter is a hot body such as the sun and direct sunlight has the reasonably balanced spectrum of white light. Light used for general illumination is white, but there are many types of white light, each having different spectral power outputs.

Lighting contributes to expectations in four ways:

It provides a framework within which expectations are generated and getting the illumination correct is fundamental to the creation of an atmosphere within which we sit down to eat or do our shopping.

It forms the link between the customer and the product being sold.

It can render and change the appearance, that is, colors and shapes, of the product itself.

It possesses an appearance and provides an atmosphere in which we can go about our business. Feelings generated within us change with illumination type, many finding that, for example, daylight is preferable for the work environment.

Hence, within the physical design as a whole the skilled lighting designer has the power to manipulate the product, the ambience, and hence the expectations of customers through the generation of mood.

LIGHTING OF ROOMS AND SPACES

Lighting is a building tool that provides a working and living environment. Outside, for the law abiding, light creates an expectation of bodily security; for the thief, it creates one of insecurity. Light communicates; it invites people in and controls circulation through attracting individuals to areas of brightness. It creates visual comfort and atmosphere, conveys information; it sets off objects and contributes to the welcome received. Lighting divides spaces into areas of light in which activity is encouraged and surveillance effective, and areas of shadow that cultivate peace, calm, theft, and immodest behavior. This applies to the restaurant, store, and home. Lighting and color are used to structure spaces and may create a sense of continuity or difference between adjacent areas. Where people and food are involved the color-rendering quality of the lighting needs to be chosen with care. It is easy to reduce expectations of food and to make people appear less desirable merely by using the wrong light source.

Stores that receive part of their interior light from daylight are felt to be more comfortable than those employing only fluorescent. Tungsten and candle light create a more intimate and warmer atmosphere than fluorescent. Although artificial daylight fluorescents are recommended for stringent color matching work they provide an uncomfortable working environment.

We have evolved to regard the human complexion illuminated by normal daylight as desirable, hence lamp manufacturers are continually searching for artificial light sources that improve color rendering of skin tones. The ideal lamp would be one that makes both people and food look ideal. This is needed for food stores where both people and a wide variety of foods are gathered together. Unfortunately, the ideal has not been achieved, but reasonably effective lighting systems are available for home and food display.

A combination of adequate intensity coupled with good color rendering helps provide a suitable environment for attracting and keeping hold of the customer. Lighting contributes to and glare detracts from feelings of visual comfort. Glare occurs when one part of the environment is much brighter than the level of the surrounding brightness. Eye muscles are forced to work constantly to compensate and this induces headaches, stress, and tiredness. Direct glare occurs when a bright light source is seen against a dark background. Indirect glare is caused by reflections from light colored surfaces. Hence, light should be sufficiently intense and angled so that small labels can be read and food identified or examined, but not so intense and angled that it creates glare and visual discomfort.

Lights of different color have different appetite appeal and the wrong lighting for a dinner table can ruin the most delicious meal. If the illumination contains too great an emphasis of shorter (bluer, cooler) wavelengths, or has a very reduced spectral output (such as mercury or sodium vapour lamps), the color of the food as well as our companions will be far from optimum. Having got the color balance of the illumination correct, that is, so that the most desirable red-orange and orange foods are seen to advantage, almost any intensity of illumination is tolerable and can be adapted to by the visual system. Many large restaurants use versatile lighting installations. A bright flood of light can be used to encourage a brisk lunchtime trade, while a softer atmosphere of candle-like warmth is used for a longer and more relaxing evening meal. However, customers who have been working all morning in a brightly lit area may appreciate a warmer, darker, quieter decor at lunchtime. In the intimate restaurant, unlike the store, we are more likely to need the intimacy of isolation in our own separate world.

Chain restaurants such as Macdonald's and Burger King are so boldly red and brightly illuminated that there is no doubt in the mind of diners that lingering over and after eating is not welcomed by the management (Green-Armytage's *bright* or *vivid* reds; see Chapter 2). On the other hand this appearance style creates the expectation that a cheap meal is available but that profits lie in rapid turnover. It is common that the more subtle the lighting and sophisticated the decor the greater the cost of the experience. On the other hand, lingering on the part of the diner is tolerated, even welcomed, in that there is time for the purchase of more food and drink. Hence lighting and color are a code giving a good clue to likely cost. But beware; even in some high-cost venues it is easy to outstay one's welcome as the fortune-seeking proprietor aims to double the evening's profits by serving a second group of customers. Practices vary internationally; for example, in Italy it is common even for expensive restaurants to be well lit. There, clear sight of the food increases the sensory pleasure of the meal.

For the home and intimate restaurant the less efficient but very effective tungsten and rich warm fluorescent lighting is generally found. Good color-rendering lighting is called for in large eating areas such as canteens. However, often the effect is spoiled through the poor maintenance practice of replacing tubes of a different color. It is a sign of lack of care, one that might be transferred to the food produced there.

Lighting is designed not merely to illuminate but to seduce the customer into buying. Good stores have a long history of attention to lighting. By 1914, J. J. Sainsbury had replaced gas lighting with electric in most of his stores and, when he died in 1928, his last words were said to be "Keep the shops well lit." Soon after World War II Sainsbury's were among the first to combine the new fluorescent lamps (then bluish) with tungsten (reddish) supplementary lighting. This produced a fairly balanced white light (Williams, 1994).

There are two major technical considerations in the lighting of food products. Heat generated by light fittings and lamps must be kept to a minimum and, because appearance is a key factor in purchasing, color-rendering properties of the light produced must be good.

Lighting adds drama and attracts attention so should not be uniform and boring. We are attracted to brightness and pass by areas of dimness in which products remain invisible, but it is the merchandise rather than the aisle that needs the lighting. However, we need to see a limit to the extent of a store. It should not appear to stretch into infinity, making the shopping trip seem long and never ending. Hence, unshelved walls can be marked and given a clean look with wall washing uplighters. Good lighting from above illuminates products and makes them visible. A common form of store illumination consists of recessed ceiling downlighters with adjustable angled dichroid tungsten lamps. This combination gives good horizontal illumination of shelves and displays. Concentrated banks of downlighting can be used for illumination of counters. Light colored walls and ceilings bring another opportunity to display a clean visual image and bring extra light to the scene. Energy costs are leading to a reduction of light intensities in stores. Lower levels along aisles tend to draw attention to more brightly lit shelves and reduce the intensities needed to spotlight special areas.

Practice differs widely from store to store and lighting may be modified for different areas of the store. Fruit and vegetables may be lit with blue to boost color and freshness with the more attractive fruits stacked near the entrance and highlighted with spotlights. Lighting of jams, tins, and packet foods may be more subdued and softer to encourage browsing.

Backlighting may be used in the drinks section to show off the glass, transparent reds, and clear products.

High color temperature metal halide lamps are better at lighting dark colors such as blues and greens. Tungsten halogen lamps give good color resolution to softer colors such as reds and yellows and tungsten point sources are used to create shadow as well as highlights. The lighting is, for good or bad, part of the brand and many stores have lighting circuits for both daylight and night time (Turner, 1998)

Although the pot of paint is of uniform color, the color on the wall is elastic. It varies according to the light and shadow that fall onto the wall and according to the colors adjacent to it (see Chapter 8). A yellow paint might appear greenish-yellow under fluorescent lighting, but look strong and slightly reddish under tungsten illumination (Billger, 1999). In sunlight, all colors become more chromatic; yellow colors become 'warmer' and more yellowish, greenish and bluish colors shift towards green, while reddish-blues behave less predictably. In skylight, a pale yellow may become greenish and increase in grayness, while a blue color may become more bluish in hue and chroma (Hårleman, 2001). Outdoor studies in Sweden show consistent differences between the inherent color of the paint in the pot and the perceived color of the building façade. Reddish-yellows tend towards yellow and greenish-yellow, green yellow colors tend to increase in greenness and blue green while blue greens tend towards blue. Yellowish-red colors tend to move towards red and reddish-blue (Fridell Anter, 2000).

Lighting is fully exploited in the theater, where position, intensity, diffusion size, and form of lighting can be varied. Here it sets the realism, that is the time and season, of the scene as well as contributing to the dramatic tension of the script and the psychological or social situation of the plot. Alternatively lighting is used to obtain a plastic effect (Rinaldi, 1998). Out of doors, yellows freshen the country scene, seductive red yellows, sinister blues, passionate purples, and bloody reds are used to support dramatic emphasis. With care some of these effects can be used in the food environment.

Exteriors can be illuminated with lighting of different colors. Mercury lamps yield bluish light, mercury halide bluish-green, sodium lamps yellowish-orange, filament lamps reddish or yellowish and fluorescent lamps bluish, greenish, yellowish, or reddish.

LIGHTING OF FOODS

In the food business the color-rendering properties of lighting can be used in different ways. They contribute to the visual quality of the general environment but they can be used in display to conceal low quality in products. Lighting is used for food displays, but light also induces photochemical reactions in foods.

Food Displays

We need to be able to see our food to check its quality. We are often in a food selection situation in which we do not trust anyone else to make the choice, but this is not always the case. Some stores use lighting that renders food color poorly and food displayed there does not look at its best. In order to sell, these stores must have a high, unblemished reputation for the quality of the food they offer. Similarly, if we trust the cook in the restaurant we can eat our meal under such low levels of illumination that most color-based means of quality recognition are ineffective.

Most foods suffer when illuminated by lamps that render color poorly but no one lamp is ideal for all foods. Incandescent lamps render colors well but have been generally abandoned for illumination of food displays because they are inefficient and emit too much heat. Directional lamps bring out sparkle desirable for products such as fresh fish and vegetables. Surface texture is also an important visual feature and this can be emphasised by using angled lighting. However, glossy surfaces, incorporated into stores because they are easier to keep clean, should not appear to be glossy when looked at by the customer. The slightest piece of dirt on such a surface will be obvious, but the presence of glossy areas can generate glare and distract the eye from the food displayed. These factors result in the use of the complex lighting arrangements encountered in many good quality stores. Good color-rendering lamps should be used irrespective of the food being illuminated.

Good color-rendering white light should be used for meat and meat products. The customer selects meat on the basis of its attractiveness and the rate of sale is directly related to the degree of discoloration. The shopper discriminates against the presence of 10 to 20% of the brown metmyoglobin pigment. However, red light hides the presence of the brown oxidised meat pigment metmyoglobin. Although such meat can be perfectly wholesome, it is not seen to be so by the customer. Hence, meats that may be rejected when illuminated by lamps used for critical color examination may be acceptable under red-enhanced illumination. However, injudicious use of red-biased lamps causes fat to appear pinker, thereby reducing the contrast between meat and fat. Some reddish lamps render beef fat an undesirable yellowish color. It also renders chicken flesh pink; this may lead to rejection. The use of even redder lighting results in the color of fresh beef becoming uncomfortably vivid. This practise is widespread. Unfortunate effects occur with processed meat also and poor color rendering can make bacon purplish or greenish.

The bright lights of the carvery unit add to the appeal of beef, lamb, and pork, as well as vegetables. However, they can detract from the quality of appearance of steam-cooked chicken.

Slightly red-biased light makes bread and baked products appear appetising but yellow-biased lights make them look as though they contain synthetic dyes. Delicatessen foods are usually subject to relatively intense scrutiny by the customer so they should be lit more intensely. Vegetables often benefit from green-biased lighting but this must be discrete because it will make the customers themselves appear totally unacceptable. Low temperature spot lamps are used to enhance the appeal of fish, fruit, and vegetables.

There are still large stores where unsophisticated lighting prevails. A supermarket in Perth, Western Australia, is an example where lamps of grossly different color illuminate each section. The customer moves from bright blue sky-light into a comparatively low intensity white fluorescent light gloom. At the end of the first aisle is heavily red-biased illuminated over-red beef on a 'white' background that is rendered bright pink. On turning right to the delicatessen counter, ham and luncheon meat dance a fluorescent pink under warm white light. On turn-ing again and while the eye is still accommodated to the dancing pinks, green vegetables look dead, even against a black background under the cool white tubes. There had been no coordination, no subtlety, and no looking at the goods from the eye of the customer.

There is no method for numerically rating 'honesty' for light sources that define the extent to which they bring out the natural color of the product.

There are no regulations concerning in-store use of unsuitable flatter or glamour lamps.

The good lighting designer will ensure that each product group is lit to its best advantage while making the store as a whole appear a unity, and avoiding the need for continual changes in adaptation as the customer walks around.

Photochemical Reactions

If all foods were wrapped in transparent film we could examine each before making a purchase. The reason frozen packs are wrapped in opaque film is that we would see ice deposited inside as a result of fluctuating storage temperature. This detracts from the appearance of the beautiful bright green peas we have in our mind when making the purchase. Another reason for the opaque wrap is that many plant and meat pigments are sensitive to light, so unwanted photochemical effects can be avoided by keeping such foods in the dark. Ultraviolet wavelengths are present in most light sources and high-energy ultraviolet and visible radiation cause color and flavor deterioration in many products. Sunlight, even when filtered through window glass, contains such radiation as do common white fluorescent lamps, germicidal lamps used in cold storage areas, and 'black' lights used to detect the presence of some types of contamination.

Long storage foods such as fruit and vegetable juices are stored in opaque packs. Most transparent wrapping materials and containers transmit light but reduction in light intensity limits damage to the product. Exposure to light of foods such as fresh, frozen, and cured meats must be limited. Problems have increased through the practice of packaging small portions of many products in transparent film. Many items on open display on the delicatessen counter rapidly degrade in color and the period of display of, for example, surface slices of cured meats must be restricted.

Many fruit and vegetable systems are sensitive to light; green toxins readily form on the surface of potatoes so stored. This does not degrade on cooking and should be cut out before use. Degradation of chlorophylls and carotenoids in vegetables, and anthocyanin pigments in fruit also occur after processing or bottling if exposed to light.

Color changes occur in beer, wine, and high-alcohol products on exposure to fluorescent and natural light. This is the reason why for these and other products brown bottles are preferred to those made from colorless clear glass. Any reduction of light entering through the glass reduces the degree of degradation. Light is responsible for changes in bouquet and taste as well as color, and clear glass should be reserved for fast-turnover short-life brands.

LIGHT SOURCE QUALITY SPECIFICATION

Light source quality can be discussed in terms of *color temperature*, the ability of the lamp to render colors faithfully, the efficiency with which it uses electrical power, and its useful life. When light sources are compared together it is seen that a white surface reflecting light from an incandescent tungsten lamp is redder (or warmer) than one reflecting the bluer (colder) light of day or that of white light fluorescent tubes. As noted in chapter 2, whatever white light source we become adapted to yields a white reflected color from a white surface. Illumination color is quantified by the lamp's correlated color temperature (cct) in degrees Kelvin. For a tungsten lamp the cct is 2700 K. For an artificial daylight tube it is 6500 K, the color temperature of standard daylight specified in British Standard BS950 (BS1967). The higher the cct, the bluer is the light. As noted above, some redder lamps have unfortunate effects on the color of fresh meat and its surround.

Lamps render colors to different degrees of faithfulness. *Color rendering* is the effect of an illumination on the color appearance of objects, in conscious or subconscious comparison with their color appearance under a reference illumination such as tungsten or daylight. Although colors look approximately the same under different phases of daylight and under tungsten lighting to which we are adapted, they can look very different when viewed under other lamps. An approximate measure of color rendering is available. This is based on a comparison of the size of the difference in color of a set of colors when they are illuminated by a test lamp and a reference light. The normal reference is daylight or 60-watt tungsten. An exact reproduction of the performance of the reference illumination yields a color rendering index (R_a) value of 100. A high R_a does not, unfortunately, guarantee a high degree of color rendering of a particular material, it can only be taken as a guide when evaluating a lamp for practical lighting tasks. This applies especially to food materials such as meat, greens, and fruit because of their particular reflectance character.

The rendering of an object color depends on the spectral qualities of light source, object, and visual mechanism of the viewer. If any one of these factors changes, the perceived color changes. Light sources having good color-rendering properties tend to have better *visual clarity*, which is related to the feeling of contrast existing between widely different colors (Hashimoto and Nayatani, 1994). Using such lamps, lower levels of illumination are required for the performance of visual tasks.

These measures are attempts to describe the effectiveness of practical illumination systems. Lamp efficiency, or *luminous efficacy*, is determined from the amount of visible radiation emitted per unit of electrical energy input, that is, lumens/watt. This is a valid measure of efficiency of performance but cct and R_a

are merely guides for use in specific situations. They are not firm measures of the effectiveness of a lamp for any particular viewing situation.

The color of the human face changes markedly with the light falling on it. Lamps that render the face agreeably are termed 'flatter' or 'glamor' lamps. These are useful for the social areas of hotels, restaurants, and some areas of public buildings. As has been indicated, their use in food sales areas is ethically and aesthetically doubtful.

If there is not sufficient light we cannot see what to buy so the supermarket ensures an adequate level of illumination. We might not be able to read the price label because the print is too small but we can always see the product itself. The eye works with reasonable efficiency at a light level of 200 lux, although performance of critical work might require 500 lux. In display cabinets intensity may reach 4300 lux. Food store general lighting normally encountered ranges from the superstore maximum of 1000 lux to the supermarket at 750 lux, and the specialist store at 500 lux. In current energy saving lighting schemes light is concentrated on the product rather than the aisle. An aisle intensity at the shopping basket level of 180 lux permits lower than normal intensities to be used on the product.

PRACTICAL LIGHT SOURCES

Lighting systems consist of the lamp and its fitting. In broad terms the lamp and filter generate the color and intensity and the fitting determines the size, shape, and drama of the illuminated scene. Practical light sources are of three types, incandescent, discharge, and fluorescent. Tungsten *incandescent lamp* radiation arises solely from heat and operating within a color temperature range 2500–3000 K. Color definition of reds and oranges are strong, but blues are dulled. Life is restricted because the filament metal evaporates and is deposited inside the glass envelope. The tungsten halogen lamp has a silica case, which permits a higher running temperature and contains halogen gas, which prevents deposit formation. Halogen lamps are more efficient, have a longer life, and, because of the raised color temperature, render blues better. Tungsten lamps are useful in retail display as they have good color-rendering properties, are flattering to human skin tones and easily controlled. However, they run hot, only 5% of the emitted energy is in the form of light, and the lamps have relatively short lives.

Discharge lamps depend on bombardment of atoms of a vapour with electrons from an electric current. Narrow wavelength bands of radiation are emitted, so colors illuminated by them lend to be severely distorted. There are four types of discharge lamp, the low-pressure sodium, high-pressure sodium, mercury, and metal halide lamps. Compared with incandescent lamps they run cooler, give a greater light output, and are more efficient. The low-pressure sodium lamp is one of the most efficient light sources, but, because it renders color very badly, is normally used only for orange-yellow street lighting. The output of high-pressure sodium lamps is broader and its color rendering is sufficiently good for sports halls and general external lighting in shopping areas. High-pressure mercury-based lamps common in industrial interiors and warehouses emit a bluish light having poor color rendering. This is improved when the inside of the lamp case is coated with a red-emitting phosphor. Addition of metal halides to the mercury considerably improves color rendering. These metal halide lamps have a higher efficiency, greater lumen maintenance, and can have good color-rendering properties. They are a suitable substitute for daylight and useful when used with low natural daylight levels, although some can have a cold appearance at night. High-pressure xenon arc lamps have a spectral output close to average daylight, hence its use for floodlighting.

Fluorescent lamps are gas discharge tubes that have a coating of fluorescent phosphor on the inside of the glass case. Ultraviolet radiation absorbed by the phosphor is re-emitted as visible light. The many phosphors developed have led to a number of tubes emitting white light of varying approximations to daylight. They can be obtained in a range of sizes, have a longer life, and are three to four times more efficient than tungsten halogen lamps. Standard fluorescent tubes incorporate conventional halophosphate phosphors. The first triphosphor fluor-

TABLE 3.1. Performance Figures of Some Commercial Lamps Found in Food Stores (from Manufacturers' Literature)

	Correlated Color Temperature (cct) K	Color Rendering Index (R_a)	Luminous Efficacy (lumens/watt)	Life (hours)
Tungsten filament	2650–3400	100	10–20	1000
Tungsten halogen	2800–3100	100	13–30[a]	3000–5000
Low-pressure sodium	1800	0	100–200	16,000
High-pressure sodium	2000	25–60	60–150	15,000–55,000
Mercury (color corrected)	3800	45	20	12,000–25,000
Metal halide	3000–6000	65–93	70–100	3500–20,000
Fluorescent	2700–6500	50–95	40–100	5000–15,000
Food retail display lamps[b]	3300–3800	75–92	37–49	5000–15000

[a]Many tungsten halogen lamps are fitted with cone reflectors and lenses; hence, light output is normally quoted in terms of beam spread and peak intensity at a standard distance.
[b]These fluorescent lamps are designed to enhance and differentiate colors of meat and other foods. However, the spectral distributions of some of them are too dissimilar from white light; hence, cct and R_a ought not to be calculated.

escent, or prime color lamps, utilised narrow-band phosphors emitting discrete spectral lines in addition to the light emitted by the halophosphate phosphor. In modern three-band tubes, however, the halophosphate phosphor has been dispensed with. These offer relatively high light outputs, lower energy costs, and good color-rendering properties; hence their wide use in retail stores. Their lumen maintenance is also better, retaining 95% of their initial lumen output after 8000 hours of burning as against 70–75% for standard fluorescent tubes. However, some colors are distorted and these tubes are not suitable for critical color evaluation. The color performance of the five-band tube is better, but efficacy is reduced. Red-biased tubes are sold for illumination of fresh meat and bread, but some have such distorted spectral outputs that the standard color-rendering determinations ought not to be made. Typical ranges of luminous efficacy, R_a, and correlated color temperature for a number of commercially available lamps encountered in food stores are listed in Table 3.1.

Although many are not suitable for the lighting of foods, they can nevertheless be found in retail lighting installations. When the product reaches home it looks nothing like the color it was in the shop. It is no use returning the product to the store because under the original illumination it might look perfectly good.

REFERENCES

Aston, S. M., and Bellchambers, H. E., 1969, Illumination, colour rendering and visual clarity, *Lighting Res. and Tech.* **1**:259–265.

Burns, R., 2000, *Principles of Color Technology*, John Wiley and Sons, New York.

Billger, M., 1999, *Colour in Enclosed Space*, Chalmers, Gothenburg.

Birren, F., 1969, *Light, Colour and Environment*, van Nostrand Reinhold, New York.

BS 1967, British Standard specification for artificial daylight for the assessment of colour, Part 1, Illuminant for colour matching and colour appraisal, BS950, part 1.

Danger, E. P., 1987, *The Colour Handbook*, Gower Technical Press, Aldershot.

Fridell Anter, K., 2000, *What Colour is the Red House?* Institute of Architecture, Royal Institute of Technology, Stockholm.

Hårleman, M., 2001, Colour appearance in rooms lit by daylight, *Nordisk Arkitekturforskning* **14**:41–48.

Hashimoto, K., and Nayatani Y., 1994, Visual clarity and feeling of contrast, *Colour Res. Appl* **19**:171–185.

Hutchings, J. B., 1999, *Food Colour and Appearance*, 2nd ed., Aspen, Gaithersburg, MD.

Kropf, D. H., and Hunt, M. C., 1984, Effect of display conditions on meat products, in: *Meat Industry Research Conference Proceedings*, Washington, American Meat Institute, 158–176.

Morgenstein, M., and Strongin, H., 1992, *Modern Retailing*, Prentice-Hall, New York.

Pegler, M. M., 1991, *Food Presentation and Display*, Retail Reporting Corporation, New York.

Rich, D. C., 1998, Light sources and illuminants: how to standardize retail lighting, *Textile Chemist and Colourist* **30:**8–12.

Rinaldi, M. R. A., 1998, Technical tools for aesthetic expression in stage lighting, *Proceedings of the conference 'Colour between art and science'*, Institute of Colour, Norwegian College of Art and Design, Oslo, 164–165.

Turner, J., 1998, *Designing with Light, Retail Spaces*, Roto Vison, Crans-Près-Cèligny, Switzerland.

Williams, B., 1994, *The Best Butter in the World*, Ebury Press, London.

4

Expectations and Appearance of the Stranger

EXPECTATIONS

Although we go to the shop or restaurant to purchase or to eat, we are also faced with other human beings each presenting their own individual, collective, or corporate appearance images. The manner of the observation and subsequent interactions lead to the generation of personal expectations and images.

Our expectations of an environment may already have been tainted by the manner with which we were treated on the telephone. An unfortunate tone of voice cannot be compensated for with a smile as it can when strangers meet face to face. As a stranger, our image of a space or person is highly sensitive. We search for any clue to quality. The initial image depends critically on the welcome or nonwelcome greeting whether it is on the telephone or in person.

We all play a number of roles in our daily lives, perhaps parent, employee, and cyclist. However, in this chapter only the role of a stranger is considered, the stranger any of us might meet in any environment. This stranger could be the store shelf stacker, cashier, part-time bar person, head waiter, chef, fellow shopper, or diner, or the stranger who is there to satisfy our purchasing needs and who represents or has to answer for perceived shortcomings of the organisation for which he or she works. It could be a stranger into whose trolley we have crashed, or whose beer we have accidentally spilled.

Social interaction takes place within a cultural setting. This setting encompasses language, specific behavior patterns relevant to specific human contact situations, specific moral values occurring within the contact environment of technology, material culture, and current events. Our attitudes can be described mainly in terms of two scales: dominant to submissive and friendly to hostile. Before we make contact with a stranger we use observations of nonverbal behavior and arrive at a judgement using combinations of these scales. For example, the pub bouncer appears to be dominant, but neither friendly nor hostile. Men and women judge traits differently. Women tend to observe personality traits and social style, men concentrate on status and achievement. People are sensitive to what concerns them. For example, some treat people of a certain social class differently from members of another, some judge on perceived intelligence or ethnic background (Argyle, 1990).

As we face, or contemplate facing, a situation involving another human being, our initial attitudes are set by our own viewer-dependent factors (see Chapter 1), in other words, by the current state of our receptor mechanisms (e.g., are we wearing our spectacles, or are we being seduced by a wonderful coffee aroma), by our inherited and learned responses to specific objects and events (e.g., by our upbringing and experience), and by our immediate environment (e.g., are we stressed by crowding, climate, or a medical condition). Such factors set the scene as to how we are going to cope with our forthcoming contact with the total appearance of the stranger. First stereotype impressions are formulated from the look of the stranger, their dress, physical features, and body language. As we treat human beings in different ways, it is necessary for us to categorize them quickly according to age, sex, and division, such as class, occupation, and physical characteristics.

In the context of our presence in a scene, our behavior is governed by two factors, first, according to our inner drive, that is, how we want to behave in response to a particular mental state, which in turn is affected by events going on around us. Second, we behave according to how we expect onlookers will feel on witnessing our behavior. These two factors need not occur in equal proportions (Ajzen and Fishbein, 1980). For example, the link between heavy alcohol consumption and air rage results in a lowering of the second factor, what the audience feel, in favor of nonsocial behavior the perpetrator subsequently regrets—after seeing the reaction from his peers on sobering up.

Similarly, trolley rage occurs when stores are crowded, causing shoppers to lose their temper as progress is slower than anticipated. The inner drive takes over and the unruly behavior resulting occurs irrespective of what others in the vicinity may feel. In most circumstances, however, the majority of the population behave in a perfectly reasonable manner, expecting and attempting to go about their shopping and eating with minimal conflict and stress.

Using the general model for the perception of a scene, (see Chapter 1) we derive expectations of the stranger from their total appearance. These expectations can be divided into five parts as indicated in Fig. 4.1. Each of these five groups of expectations is discussed in turn.

VISUALLY ASSESSED SECURITY

When faced with a stranger, visually perceived cues give rise to expectations of the potential outcome of the encounter. Three types of security expectation

Visually assessed safety.
Visual identification of the stranger(s)
    Visually assessed recognition of person 'type' and associations with respect to previous
        encounters with this 'type' of person
    Visually assessed convenience, e.g., must I engage with and do I have time to engage
        with this stranger?
Visually assessed usefulness of the stranger(s) relevant to the scene
Visually assessed pleasantness of the stranger(s) relevant to the scene
Visually assessed satisfaction of the outcome from the proposed interaction with the
    stranger(s)

FIG. 4.1. Expectations resulting from stranger/stranger interactions

arise in all scenes (Fig. 4.2). The nonverbalised question we might ask ourselves
when entering a space is "Will I, by entering this place and consorting with staff
or customers, be led into physical danger?" The extent of possible physical
danger varies considerably. A confrontation in a pub may form an amusing
diversion because there is space to escape from the action. In an aircraft, however,
flying at 30,000 feet, a thoughtless confrontation could lead to the death of 300
people. The variety of human life being what it is, some individuals go to a
particular venue specifically to engage in combat. The questions these people ask
of themselves are obviously different from those asked by the bulk of the
population. What clues are there to the potential outcome of stranger contact?

    Communication may be verbal or silent. When we interact with close friends
it is often sufficient just to listen to what they say. When we interact with
strangers, or near strangers, we may need to absorb information from as many
sources as possible to get the complete message the stranger is conveying. These
sources include body language, of which there are three types, as listed in
Fig. 4.3.

    If we have sufficient information we can make an analytical assessment of
someone's trustworthiness. The amygdala, a region of the brain controlling our
fear response, also appears to match our first impressions of people with our

Visually assessed security of body, e.g., "Does the stranger pose a physical threat?" or "Is
    this person a physical bully?"
Visually assessed security of mind, e.g., if I am a calm individual, I will look for answers to
    the questions: "Will I be able to communicate with this person?" or "Does this person
    appear to be a verbal bully?" or "Will I be in control of this situation?" or "Am I
    performing within my comfort zone?" Alternatively, if I am feeling angered or assertive,
    I may think: "Will I be able to bully this person?" or "Can I take him/her out of their
    comfort zone?" or "Does this person present a sufficient challenge to me in my present
    angry mood?"
Visually assessed security of spirit, e.g., members of some religious sects will not sit down
    to eat with a nonmember for fear of having their spiritual purity compromised.

FIG. 4.2. Visually assessed security types.

Body language related to aggression, or the threat of potential aggression.
Body language related to nonaggressive verbal communication.
Body language related to nonverbal, silent communication.

FIG. 4.3. Types of body language.

knowledge of past experiences. This enables us to make judgements of trust-worthiness in the absence of other information (Adolphs *et al.*, 1998).

While we are asking the supermarket assistant where the breakfast cereal is located we do not expect to be greeted with verbal aggression, obscene gestures, or the threat of actual bodily harm. This may happen when the store is full at Christmas time, when aisle widths have been reduced with seasonal goods and shopper comfort sacrificed. At these times, competition between customers for trolley space can be fierce. Requests such as, "Please excuse me while I reach across for the pack of coffee," may not be greeted in the friendly manner expected at other times of year. This type of shopping experience does not provide a good foundation for the festive season.

Visually assessed security is judged from three aspects of the stranger's body language: physical appearance, personal space, and nonverbal communication. Each aspect is discussed in turn.

Physical Appearance

Among the most immediate physical appearance cues are stature, physical attractiveness, facial expression, pupil size, and bodily activity. There may be five kinds of inference process accounting for the conclusions drawn about a person from the sight of their physical characteristics. *Temporal extension* results from a momentary expression; for example, a smiling person indicates someone who is good tempered. *Parataxis* is a generalisation from the characteristics of someone we know; for example, if your mother was warm and kind you might infer that a woman who looks like your mother is also warm and kind. *Categorisation* is stereotyping. This occurs, for example, when we categorize someone as a member of a group, such as based on race, sex, or age. *Functional quality* is based on the functions of specific parts of the face; for example, glasses may indicate someone who does much reading, and is therefore intelligent, while thin-lipped people are 'tight-lipped' in personality. *Metaphorical generalisation* is a more abstract generalisation. For example, a thin-lipped woman who fashions her hair into a tight bun may be equated with someone who is severe and who is, therefore, censorious (Hinton, 1993).

There are many examples of beliefs about the way we look. Wearing of glasses makes no difference to facial attractiveness but give a person greater perceived authority. Hairstyle is influenced by fashion, cultural factors, group membership, and social group. Men and women tend to equate red hair in a woman with fieriness and aggressiveness. Men see blonde women as less intelligent and less aggressive, but women do not. Both sexes agree that blonde women are more popular than brunettes. Adults who have baby faces are often accorded the characteristics of a young child, that is, warm but lacking in physical and social power. It is inferred that overweight people are lazy and greedy, that taller men are more competent and more desirable, that those with scarred faces are more dishonest and less sincere, that someone who speaks with a regional accent is of lower status, and that those having long hair are less business-like. There may be little or no evidence that these associations are true but the above five rules explain the mechanisms of these widely held beliefs. Physically likeable people tend to be more persuasive. Similarly, advertisements featuring celebrities, or those that are likeable, or those from the audience to which we ourselves belong, tend to be more believed.

The attitude of the stranger may also be judged by body activity. When sitting, a relaxed attitude can be communicated by an increase in the angle at which the body reclines, by relaxation of hands and neck, and by asymmetry of arm and leg positions. When activity increases, speech rate, body movement, gesture rates, and facial expressiveness increase. Any further activity increase may be interpreted as an aggression signal.

In the United Kingdom, the handshake is a conventional greeting when being introduced or on the occasion of a more formal meeting. It has been the rule that, after the initial handshake, two people conversing do not touch each other. This convention is now becoming relaxed. However, the area brushed should go no further than the point of the elbow and certainly should last for no longer than a very few seconds. Otherwise an inappropriate interpretation may be placed on it.

Appearance of serving staff is of particular interest for restaurant managers as well as customers. Although kitchen staff may be permitted a certain freedom as regards dress, the grooming code applied to serving staff is more stringent. In general, restaurant managers are strict about the appearance of fingernails, the wearing of scented goods, the amount of personal jewelery on display, and the wearing of polished shoes of good condition. Hair is a hygiene issue and some restaurateurs ban dishevelled and braided hair, and insist that long hair is secured with a clip, band, or ribbon. Moustaches and beards may be banned altogether. Staff that are tidy, smart, unobtrusive, look conventional, and fit in with the surroundings are a welcome sight to the customer. Uniforms are discussed in the section on visual identification.

We act as individuals or we act as a member of a group behaving according to the nature of the ideology of the group. For example, as shoppers we may arrive at the store as individuals, but when we are inside we act as shoppers are expected to act. For example, we collect a basket at the door and proceed at a pace and along a route convenient to enable the shopper group to behave efficiently; we then queue up quietly and wait for the cashier to attend to us. This is an historical agreement of behavior suited to our best collective interests. The bigger the crowd the more patience and consideration we must show. If, however, something goes wrong, say during the pre-Christmas rush, and tempers become frayed, the nature of the crowd could change. At sale times at large stores the confrontational attitude sometimes surfaces, and fights may break out as individuals strive to complete their shopping. In other words, one historical behavior pattern breaks down in favor of a second.

Personal Space

Invasion of personal space may be interpreted as aggression. We are able to tolerate the presence of a stranger within a distance of approximately 1.2 m (4 ft). For family and friends, as well as waiters, a distance of 45 cm ($1\frac{1}{2}$ ft) is tolerated. More formal exchanges take place within the social zone between 1.2 and 3.7 m (4–12 ft). Personal space requirements depend on culture, race, sex, degree of familiarity and social position. Arabs tolerate smaller spaces, those of Mediterranean and Middle Eastern cultures larger, and North Americans and northwestern Europeans larger still. Interacting males need a larger space than females. Invasion of personal space can be interpreted as aggression, or it may express a desire for intimacy. Interaction during service in a shop is normally governed by the presence of a counter, which acts as a half-way position preventing violation of the space.

Sometimes we cannot help invasion of this personal space. When crossing the room, even though we may not look directly at others, we are aware of their presence and make adjustments to our route to avoid collision. In the store, when such adjustments are impossible and no movement can occur, we wait patiently for a gap to appear or we engage in trolley rage and attempt to barge others out of the way. When crammed into a congested train, passengers pretend others do not exist by looking or leaning away, eye contact is avoided and we stare into the middle distance. People may remain still, avoiding the potential for communication. In doing this we consciously avoid making verbal contact while remaining visually aware of potential danger.

Self awareness arises when we are alone with someone. However, crowds have little sense of self and being part of a crowd can be stimulating. Walking into a crowded public house, we apologize when we become aware that we are breaking into the personal space of others, going around groups in conversation. Eating in a crowded dining room can pose problems. We can tolerate someone sitting closely beside us at the table, but the table must be of sufficient width to prevent feelings of personal space invasion towards those sitting opposite. This is because it is the distance from the face or eyes that is critical. Our personal space extending forwards is reduced when we turn our head to the side.

Public space can also be 'owned.' This occurs, for example, when the crowd of regular public house customers cluster around the bar preventing strangers from being served. Ownership of this nominally public space is communicated with barrier signals. Strangers are given icy glares, or the 'owners' remain silent and turn their backs on the newcomer. On the other hand, there may be a smile and a gap found for the stranger to occupy. This gap may well be within the personal space of each individual.

A coat draped over the back of a vacant chair may mark territorial space in a buffet restaurant. The extension of personal space to that occupied by our possessions is a possible area of conflict, particularly if the space is crowded. When we sit down in the restaurant the table becomes part of our personal space. The waiter, who may own the establishment, may be treated as an interloper and an invader of the space. On other occasions the waiter, feigning disinterest, may be ignored even when all parties are aware of each other's presence.

Wherever we are we can adapt our personal space to survive within most environments, even though we may feel uncomfortable doing so.

Nonverbal Communication

We reveal our thoughts and intentions through nonverbal communication using facial expressions, gesture, body position, distance, voice quality, and dress. Dress we can closely control, but during personal interaction we learn to control more involuntary actions with different degrees of success. However, on most occasions we attempt to present ourselves in a desirable and credible light.

We use nonverbal communication to encourage or discourage social contact, attracting attention by eye contact, or avoiding it by looking into the middle distance. This discourages conversation while we are aware of what is going on around us.

Behavior may not be quite as rigid in the food environment. Whether stranger/stranger eye contact is interpreted as friendly or threatening depends partially on the privacy of the space. Where strangers have some common focus as a group, perhaps a dinner party or a restaurant, or a party to which they have been invited, or strangers meeting in a public house, or attending an exhibition on a specific theme, occasional eye and verbal contact are not threatening. Indeed, in these cases, eye contact is a signal to initiate a conversation. However, in more potentially threatening spaces, such as the underground or a large town pub, such a gaze may be interpreted as threatening. During conversation it is natural and normal to make eye contact from time to time. However, the situation is not straightforward as eye contact behavior is culturally dependent.

In the restaurant an established system of signals between customer and waiter leads to a minimisation of meal interruption and maximisation of the diner's needs. An example is the signal that we have finished eating and therefore ready for the attentions of the waiting staff. In Britain the fork and knife are placed together in parallel on the plate. In the United States the signal may be the fork and knife placed apart at an angle.

We can take appropriate action when faced with intention movements, such as a pointed finger, a shaken fist, a furious face, or a raised arm. Our judgements of someone portraying one of these extreme feelings, such as boredom, enthusiasm, an active interest, frustration, and nervousness, are likely to be reasonably accurate. However, many less obvious examples of unintentional movements are mentioned in the literature. Not many of us are expert at interpreting the significance of such nonverbal leakage and women seem to be better at it than men.

There are many bodily movements signalling the erection of barriers and gestures of openness or congruence. For these and more subtle gestures of, for example, telling lies, it is preferable to seek confirmation before condemning the speaker. We rely on our experience and instincts but amateur analysis of the personal behavior of strangers may easily lead to erroneous judgements. For example:

> Contrast the interpretation of the gentle smile that admits guilt and the one that indicates "You have made a mistake, I have got you now."
> Contrast the shrug that indicates "I have made a mistake," with the one that says "I couldn't care less."
> Contrast the reason we cross our legs towards a companion indicating we are in synchrony with our companion with that indicating that it is more comfortable sitting that way.
> Contrast the way we wriggle in the chair indicating we are lying, with that indicating we are embarrassed, or that we may need to go to the lavatory.
> Contrast the reticence that indicates shyness, with that indicating coyness, with that indicating surliness and truculence.

Mistakes can be made because of different cultural attitudes and customs. Afro-Caribbeans may have been taught that making eye contact is disrespectful; in Western white culture avoidance of eye contact is indicative of a 'shifty' character. Confusion can also be caused through different attitudes to personal space. Cross-cultural misunderstandings lead to rejection on the grounds of not conforming to the standards of a civilised society. Mistakes can be costly. Murder has resulted from a road rage incident when a hand signal meant to say 'sorry' was misinterpreted as a sign of aggression. Only with increased acquaintance with the individual can we become more certain of their intentions.

Nonverbal Communication and Staff Behavior

Members of staff are the company's front line troops, providing the link with the customer who supplies the money to keep the business in business. It is the staff that must do the selling and they have been told to treat customers in a particular way with the ultimate aim of increasing sales. It is a major aim within the retail outlet to keep present customers satisfied so that they return on another occasion. Most customer losses occur through the indifference of one employee, only a quarter are lost through reasons of the product or competition. Dissatisfied customers tend not to complain to the business; they complain to others. Staff members behaving openly are available for communication with the customer. On the other hand, two members of staff talking together create a barrier to communication, particularly when they turn their backs on the customer. It costs up to ten times more to obtain business from a new customer than it costs to increase business from an established customer (Nash and Nash, 2000).

We can identify success or otherwise of staff training. There may be little or no overt pressure to buy—if customers are on the premises they have probably come in to buy something. In any case shoppers generally do not appreciate being pressured and it suits them when staff have been trained to be helpful, friendly, cheerful, and informed. If dissatisfaction occurs, the ability to apologize often defuses annoyance and the customer goes away satisfied. That a shopper complains is an opportunity for the management to do something about it and keep the customer as a client.

In their store training, staff ought to have been taught to use, but not overuse, the friendly, polite smile and to use facial expression to show interest as a listener. A blank face communicates boredom as well as a lack of interest. Staff ought to be told that signals of interest and concern for the customer can be expressed by leaning forward but not slumping, facing toward rather than turning away, having arms open rather than folded, using gesture, and speaking with a moderate volume with varying pitch and pace. Shoppers appreciate being treated with respect, even though they may not have much money. The policy of Wal-Mart is to treat every customer as a guest. McDonald's recipe for success is to be respectful to customers, even though it adopts a no frills approach to its business.

Greeters are becoming more common at the store door in the UK. This is a skilled work, certainly for the British, who tend to be reserved and in a hurry. Being accosted unnecessarily by a complete stranger may not be welcome. Certainly burly, scowling security staff are likely to cause a change of mind of the prospective shopper or diner, encouraging them to look elsewhere for their purchases. However, having someone around to help in times of need is appreciated. A welcome sight to the weary traveller is the hotel porter/doorman, who offers help with taxis, baggage, and advice about the hotel.

Often the bartender is the first member of the establishment to meet the customer and is in an ideal position to make the client feel welcome or merely a statistic. Like all greeters there is scope for creating tension or relaxation in the visitor. Everyone entering the bar space should be greeted within seconds with at least an acknowledgement or a smile. The bar can have different functions. It may be merely a place for drinking or a filter for the dining room. It is a focus for information and bar staff ought to know details of the running and organisation of the hotel and dining room as well as of the drinks available. Whatever the function of the bar, the good bartender will be sensitive towards the needs and feelings of customers and whether they need privacy or conversation. To make the environment comfortable for others they also have the job of identifying and evicting troublemakers.

Ideal waiting staff has knowledge, confidence, style, and speed. They also have a difficult job. Customers, otherwise quite decent people, can, when they are at table, adopt an officious or petulant or objectionable attitude. When in the presence of a stranger who is unable to answer back, customers can quickly employ a form of address more appropriate to speaking to someone of imagined social inferiority. That the diner's sugar levels may be low does not excuse this behavior. However, it is part of the waiter's skill to cope with the unhappy stranger in a professional manner and willingly do something about the cause of any complaint. From the customer's standpoint, bad behavior is silly, as serving staff, unlike store cashiers, are in a position to wreak revenge on customers without their knowledge.

Service is crucial to the meal scene setting and pleasure of eating out. Waiting staff who cause embarrassment through making customers feel they exist for the convenience of the business deserve to be politely reminded who is paying the bill.

Above all, general appearance and dress must conform to what is expected in that particular situation. This applies throughout the food business whatever the position held, including cook, waiter, bar person, store manager, cashier, or shelf-stacker—this also applies to the customer. Customers can then identify staff making an approach with confidence. In facing badly behaved customers the uniform reminds the member of staff that he or she is there to serve and that part of the job is to attempt to defuse a potentially bad situation and put right whatever problem has arisen. Surliness in other customers can usually be ignored but surliness in staff is not easily forgiven. Problems arise where there is a role ambiguity and conflict among employees, where there is mismatch of skills and technology required and supplied for the job, and where there is a misunder-standing of responsibilities for action in particular circumstances.

Buyers appreciate being greeted by someone standing straight, with a well toothbrushed smile, a firm handshake, someone maintaining eye contact, and with a friendly greeting. The ideal member of staff will proactively put things right before the customer notices and will be empowered to refund money or replace faulty goods. This initial contact is the customer's introduction to the business. The food business is competitive; it requires well-trained and knowledgeable staff of smart appearance. We customers have to be welcomed and wooed.

VISUAL IDENTIFICATION

When we make contact with other customers or employees, among questions arising are: Will we be able to communicate with this person, and will he/she be performing within their comfort zone during the communication? We use the stranger's physical characteristics and apparent social and skill characteristics. There are different motives for seeking contact with a stranger and there are three ways in which we may want to identify them.

Identification by rank. Questions include "Is this an appropriate person to help solve my problem?" or "Will he/she be able to tell me on which shelf the sugar is displayed?" or "Will he/she be able to fulfil an order for a cup of coffee?"

Identification of someone who will be sympathetic to our present state of mind. For example, I might feel sorry for myself because I have lost my wallet, or I might wish to contact other football hooligans because I feel like a fight.

Identification of someone who is similar to us. We may think that our problem will be solved better by someone of our own class, or way of dressing, or of our own sex.

Specific identification expectations relevant to contact with the stranger are included in Fig. 4.1. These are identification of the person 'type,' that is, however we personally categorize people, and convenience, that is whether we must or have the time to engage in communication.

Within the food business, the four different types of uniform have different functions. They are:

The navy blue, official-style uniform worn by security staff, for example, the police look-alike dress. This is a completely specified uniform, down to the color of shirt, tie, and shoes.

Lounge-suited security staff, the bouncer.

Lounge-suited members of management. As with the bouncer's suit, this may be an incompletely specified dress, confined to a small range of colors but with optionally colored shirt and tie.

Uniform worn by those preparing food and by waiting and serving staff. This may be a completely specified dress differing in different establishments.

Store security guards sometimes wear the official-looking uniform similar to that used by police. These are not often seen in establishments in which food and drink are consumed because they contravene a primary aim of the restaurateur, that of putting the customer at ease. Hence, lounge-suited employees are used to tackle problems of security. Many members of management also wear suits, differences in body language separating bouncer from manager. Bouncers have the potential to appear dominant and hostile, while management staff generally tends towards more subtle levels of dominance and friendliness. Uniforms of staff employed for food and drink preparation and service carry their own hierarchy, not always obvious to the customer.

Hence, as well as protecting us from the elements and the public view of our sexual signals, clothes give out messages of disposition, social rank, and group loyalties. Clothes confer beliefs about the wearer. For example, a policeman, whose uniform places him within a hierarchy, is expected to be more competent and reliable in uniform than out, and has more authority when wearing the helmet than when not. However, dress is not the whole story of appearance. Behavior as well as dress, hair cleanliness and length are important to the success with which a particular role is carried out.

Our significance as a person changes according to our dress. For wearers of more formal uniforms, certain behavior is expected as the wearer carries out the job duties customary with the rank or position indicated. Those wearing clothes associated with higher rank are generally regarded in a more respectful light. Those having ambition in a managerial sense, men slightly more than women, tend to dress in the more sober colors of gray, dark blue, tan, beige, and white, suggesting seriousness, hard work, and status. Those who do not aspire to this route tend to wear more highly colored clothes. Those expending physical effort, such as rugby players, wear wider stripes, narrower stripes being more associated with those engaged in mental or intellectual activities. When relaxing from promotion-seeking work time activities we may wear the totally informal and unspecified sweatshirt, perhaps depicting the name of pub or brewery. The wearing of a uniform identifies the wearer, and constant wearing of it can transform the wearer so that it becomes difficult for them to act normally. The wearer of the uniform relinquishes the right to act as an individual (Lurie, 1982).

In store we may make contact with a member of staff or another member of the public. There is the need to seek out and recognize a member of staff who can help to solve a specific problem, such as finding the whereabouts of the butter. For larger problems over which we feel upset we can take different approaches. For example, we might seek out a more senior management figure and discuss the problem with them. Alternatively, we might take the complaint to a junior member of staff and browbeat and bully them even though we know they are not responsible and can do little about it except to report it to a more supervisory level.

Unless we specifically wish, we need have no contact with store employee strangers. However, where food and drink service is concerned we cannot avoid it unless we are obtaining our refreshments from a vending machine. The total appearance of waiting/serving staff is most important. It can lead to outright rejection of the premises as a place to eat.

Human response to any situation depends upon many factors, for example, whether or not we are in a straightforward acquaintance conversational role, whether others or we are acting calmly or under stress, whether we like or dislike the person, whether we are sympathetic or not to their problems, whether we are feeling well or ill. Confrontational situations occurring, say, within the store where there is role play, will also depend upon whether we are, or believe we are, acting from a superior or inferior rank, and whether we are acting in a capacity of self or as an employee or, for example in a mob situation, as a collective.

VISUALLY ASSESSED PROPERTIES OF THE STRANGER RELEVANT TO THE SCENE: USEFULNESS AND PLEASANTNESS

When we have assessed a food environment for safety and have identified our food, we may assess food quality in terms of visually assessed flavor and visually assessed texture. Similarly, we mentally assess strangers in terms of visually assessed professional (or usefulness) qualities and visually assessed personal (or pleasantness) qualities. Assessments of professional qualities are made in terms of someone who has expertise, strength, interest, and profundity. Assessments of personal qualities are made in terms of one who has friendliness, humanity, honesty, sincerity, and reliability (Baggaley, 1980). Such assessments are made in terms relevant to our present problem and relevant to the present scene. As the scene changes, the terms in which we assess the stranger may change.

Using these qualities a waiter, for example, may be visually assessed for trustworthiness, sincerity, and competence. Meeting the same person in a socially equal environment, say a public house, depending on our motives for making the approach, we may visually assess them in such terms as compatibility, friendliness, discretion, and gullibility. Verbal communication, body language, and general appearance reveal to us whether we are going to have a comfortable restaurant meal or whether we want to engage others in conversation in the pub.

The significant success of Western fast food chains in East European countries is due in no small part to the way in which staff are trained to communicate with customers. Smiles replaced scowls. Other factors brought were staff who are clean, tidy, appropriately dressed, and trained to handle food and utensils correctly. Holding cutlery not by the handles or holding a glass around the rim are powerful discouragements to eating and drinking.

As a result of our reaction to and interaction with a stranger, we tend to put them into a category. This first impression may take a number of forms such as liking, loathing, or disillusionment. This leads to a desire to increase or decrease further interaction. Generally, perceived similarity with the stranger encourages interaction; differences discourage it.

The total appearance of an individual results from our deductions, using our own personal viewer-defined characteristics, from the individual's physical appearance and body language. We use our past experiences and prejudices to determine whether we are likely to come to harm at the hands of this stranger and how likely we are to gain satisfaction from contact with the stranger.

VISUALLY ASSESSED SATISFACTION

Visually assessed satisfaction is the degree of hope that the outcome of the forthcoming encounter with the stranger will be satisfactory and that the query we have to pose will be answered adequately. Within the food environment we normally approach the stranger because we have a problem. Perhaps we cannot find something in the store or we want to order a drink or to complain that there is a fly in the soup. Will the person over there be able to satisfy me? Will this person know the whereabouts of the goods I cannot find? We balance our impression of the stranger against the problem we have. Questions to be weighed include the calculation of the odds that the person will be able to answer my question or rectify my complaint. Are they likely to have the power to return my money because I am not satisfied with the goods I purchased last week?

By considering physical appearance and body language, we try to identify a stranger who is a member of staff of appropriate rank and experience. Sometimes we deliberately do not look for someone who will be able to help, that is, we need someone on whom we can take out our frustrations. Someone easily identified, who is young, inexperienced, and captive, someone who can be bullied—like the junior cashier in the cage. We remain in control of the situation while voicing our disappointments at a defenceless individual who is not in a position to fight back.

A person seen to be pleasant is seen more positively; this includes customers as well as staff. We are inclined to like someone if they like us. Someone who is pleasant, approachable, smiling, and friendly finds that others tend to behave likewise. The manager who fetches you in person or at least gets up from his seat and smiles when you enter his or her office gets the interview off to a good start. Then, if there is a complaint to deal with, the manager has not made things worse before the interview has begun.

We are more likely to visit the establishment again if we gain an image of satisfaction from friendly service or on hearing the words "thank you" or "sorry." This applies in particular to the public house. Only so much can be achieved by pub appearance and design; the ultimate success depends on the friendly face and warm personality of the person behind the bar. The rude restaurant owner can be tolerated with amusement if very high quality food is served, but to the pub, the personality of mine host is of paramount importance.

REFERENCES

Adolphs, R., Tranel, D., and Damasio, A., The human amygdala, *Nature* **393:** 470–474.
Ajzen, I., and Fishbein, M., 1980, *Understanding Attitudes and Predicting Social Behaviour*, Prentice Hall, Englewood Cliffs, NJ.
Argyle, M., 1990, *The Psychology of Interpersonal Behaviour*, Penguin, London.
Baggaley, J., 1980, *Psychology of the TV Image*, Gower, Farnborough, UK.
Clutterbuck, D., Clark, G., and Armistead, C., 1993, *Inspired Customer Service*, Kogan Page, London.
Engel, J. F., Blackwell, R. D., Miniard, P. W., 1995, *Consumer Behaviour*, Dryden Press, Fort Worth.
Hinton, P. R., 1993, *The Psychology of Interpersonal Perception*, Routledge, London.
Lurie, A., 1982, *The Language of Clothes*, Hamlyn, Feltham, UK.
Marsh, P. (Ed.), 1988, *Eye to Eye*, Guild Publishing, London.
Morris, D., 1978, *Manwatching*, Triad/Panther, St. Albans, UK.
Nash, S., and Nash, D., 2000, *Exceeding Customer Expectations*, Pathways, Oxford.
Newman, S. P., and Lonsdale, S., 1996, *Human Jungle*, Ebury Press, London.
Pease, A., 1995, *Body Language*, 2nd edn. Sheldon Press, London.
Porteous, J. D., 1977, *Environment and Behaviour*, Addison Wesley, Read, MA.
Raspberry, R. W., Fletcher Limone, L., 1986, *Effective Managerial Communication*, Kent Publishing, Boston, MA.
Warr, P. B., and Knapper, C., 1968, *The Perception of People and Events*, John Wiley, London.
Zeithaml, V. A., Parasuraman, A., and Berry, L. L., 1990, *Delivering Quality Service*, The Free Press, New York.

5

Expectations, Color and Appearance in Advertising and Packaging

EXPECTATIONS

Color and appearance provide the impact that sets up the expectations for store, restaurant, package, and advertisement. The external appearance of the restaurant or store, for example, signals not only its presence but what it has for sale, its social class, and the relative cost of its goods. The twelfth-century alewife signalled that the latest batch of beer was ready to drink by hanging a bush on the front door of her cottage. The modern pub has increased somewhat in sophistication via the permanence of its sign, the presence of an advertised name, and the colors of the exterior paintwork. Designs and colors help us carry out the daily chores and delights of shopping and refreshment with greater ease. They also assist marketers to use as much psychology or pseudopsychology as possible in parting us from our money.

Much effort is put into getting controlled amounts of colored ink deposited onto rapidly advancing sheets of paper and card. Color, when used correctly, brings specific benefits that are different according to the specific situation. The advertisement provides a stimulus designed to draw attention to a specific product, service, or venue informing whomsoever is concerned of their value. Advertisement and pack transfer information, project appropriate images, clarify identities of product and producer, evoke moods, and inform the customer about the product and its use.

Food industry advertisements work in different places:

In the home there are newspapers, direct mail, and magazines in profusion, as well as television advertising.
In the street there are facades of restaurants, pubs, and stores, as well as posters.
In the store there are packages and displays.

In all advertising media color and appearance are used to transfer information. For the product, this aims to induce temptation through invoking feelings of appetite, or craving, or tastiness. The pack usually shows a recognisable company and product logo, which says "You know me (or, you have seen me on TV), you liked me last time (or, you have been told I am nice)." It often indicates specific product qualities, such as freshness, goodness, or richness. For example, the pack may say "I have been prepared hygienically" or "I am made of pure ingredients so I will be good for you" or "I am sophisticated and you will impress your family (or friends or fellow shoppers) when they see you possess me." In other words, the pack says for the product "Come and eat me!"

Misinformation can also be transferred and pack designs can overstate product quality. Previous experience of the product will cause the customer to recognize the over-high expectations produced by the pack design. Alternatively, the design itself may lead to the belief that the product cannot possibly be of such a high quality as the pack indicates. These judgements can be transferred to other products of that type or with that logo. Wording on labels is considered in Chapter 7 under in-store irritations.

As with many other viewed scenes within the food industry both advertisements and pack arouse expectations. Following Fig. 1.1 when we look at an advertisement for food or drink specific expectations are aroused (Fig. 5.1). These expectations form an emotional description full of everyday words we all use.

Similarly, when looking at a restaurant we will identify from the appearance of the exterior whether it is likely that it will attract people who will do me harm, whether it will attract those with whom I wish to be associated, whether it is likely to be excessively noisy, whether it looks clean, and whether my personal space will be invaded unacceptably? We compare our visual identification and visually assessed relevance of the venue with the type of eating or drinking experience we require at that particular time. Also, from the total appearance we will visually assess the satisfaction we shall expect to gain from our visit.

Hence the total appearance generates expectations of a product advertised whether this appears on the page, the hoarding, or it is the exterior of a previously

---

Visually assessed safety, i.e., Do I identify with this product, will this product taste good, will it make me feel good, will it be good for my health?

Visual identification, i.e., What is this product, and can I identify it in terms of my former experiences?

Visually assessed flavor, i.e., What flavor(s) are evoked by the advertised product?

Visually assessed texture, i.e., What texture(s) are evoked by the advertised product?

Visually assessed usefulness, i.e., Is this what I need at this moment?

Visually assessed pleasantness, i.e., How pleasant will this be to eat?

Visual assessed satisfaction, i.e., Will I feel satisfied after I have eaten it?

FIG. 5.1. Expectations arising from an advertisement or pack.

unvisited restaurant. These expectations are vitally important to the feelings we actually experience when we sample the product. This chapter is concerned with advertising and packaging; building facades are considered further in Chapter 6.

COLOR AND APPEARANCE IN FOOD ADVERTISING AND PACKAGING

The driving force to create an advertisement stems from the designer's need to create an impact and induce a number of images in the mind of the reader. All scenes create within each of us an impact or gestalt image as well as more deliberate intellectual, emotional and sensory images (Fig. 5.2). Advertisement and package designers ought always to have such driving forces in the front of their minds when conceiving their creation. Through the quality of the arresting dynamic and the body of the creation the viewer forms opinions of the qualities of the brand and advertiser, the quality of the product, the quality of the advert, perhaps as a work of art or originality, as well as its comprehension or otherwise as a whole.

The first regular commercial advertising began outdoors. By the end of the nineteenth century high-speed lithographic color printing made it possible to cover walls and the insides of railway carriages with posters. At that time France was the world center of modern art and French posters were artistic creations in their own right. Private individuals were permitted to put up posters, but only in color, as black and white was reserved for official notices. During the 1880s and 1890s Mucha, Meunier, Casas, Cheret, and Toulouse-Lautrec vied for position on the billboard. Such artists were among the first to attempt to make color correspond to the image and branding of the business. They used color to give personality and depth to the brand.

Where there is an audience there is usually a message to sell. Images suitable for the location must be created in a form that seizes our immediate attention and retains it for a short but sustained period in an attempt to produce a response.

Impact images draw the attention of the onlooker to the advertisement and induce a recognition of the product and/or brand advertised.

Intellectual images, for example, induce experiences of comedy, sex, history, nature, fantasy, a kinship with others—real or imaginary—a specialist appeal (e.g., to children), product differentiation, or product similarity (i.e., copycat products).

Emotional images, for example, induce feelings of thrift, patriotism, foreign lands, beauty, exclusivity, greed, guilt, or fear.

Sensory images induce psychophysical reactions relating to color/appearance/flavor interactions, including those relating to pack identity and coding.

FIG. 5.2. Images aroused by advertisements and packs.

Whether it is a plain announcement or an imperative message, it will be designed to produce expectations and have emotional or intellectual appeal.

Brand characters were invented to promote mass-produced goods. Examples include Sunny Jim advertising wheat cereal, the lately deceased Bisto Kids and their gravy, and, later, Mr. Pickwick's cakes and Captain Birds Eye's frozen fish products. When the brand was introduced into the UK, McDonald's sales were initially low. They used a brightly colored clown to target children reckoning they would bring along the adults. These fictional characters compete with cooking sauces promoted by the pictures of real people such as Paul Newman and Lloyd Grossman. (Personalities do affect sales; comments by George Bush senior and Oprah Winfrey led to falls in the sales of broccoli and steak.)

Fictional characters help to add the visual humor that is needed, particularly in window displays which potential customers just happen to be passing. Inclusion of packets in the poster ensures that customers became familiar with the visual appearance and brand name of the product.

Brand names are the subject of much deliberation and research. They are a combination of logo, design, packaging, advertising and marketing. But they rely on human perception to make them concrete because they rely on recognition and reputation. They are a direct insight as to perceived quality and value relevant to the particular circumstance of the potential purchase or use. We all have our own visions of old established food product brand names such as Cadbury, Campbell's, McVitie's, Marmite, Toblerone, Mars, and Wrigley's. Name and visual image have remained practically unchanged since their creation. Some names are changed for pragmatic reasons. For example, the change from Jif lemon juice to Cif occured because people of some nationalities found the former difficult to pronounce. Some changes are due to political correctness. Tesco changed the name of the long established and traditional pudding *Spotted Dick* to *Spotted Richard* because they found that it embarrassed some customers. However, brand names should be changed with great care. For example, when the long established and highly regarded dairy product firm Unigate changed its name to Uniq, share value fell by one third.

In the 1920s and 1930s food product advertising scenes tended to concentrate on just a few themes:

Realism: pictures of the home or the shopping basket.

The satisfied child: who is either satisfied through eating the product or satisfied by the successful mother who is buying what the manufacturer says she ought to be buying.

The fantasy: often involving a young, good-looking woman, sometimes elegantly dressed, enjoying herself. This picture often also included modes of transport—boat, aeroplane, car, or bicycle.

The factory: food manufacturers were proud to include pictures of the large brand new factories where the product was made.

The advertiser has milliseconds to arrest the attention of the viewer, who rapidly makes a decision of whether to read on or skip the detailed message. For this reason advertisements are ever changing in spirit and design layout.

ADVERTISEMENTS VERSUS PACKS

Both newsprint publicity and packaging are advertisements. However, the mechanisms by which they become effective are different. For example, they have different constraints on size. In the case of the advert it is mainly cost. For the pack, it is product size and product type. The advert can do things the pack cannot. Within an advertisement it is easier to place the product into different and unusual contexts, content being governed purely by the designer's imagination. Such illustrations are rare on the pack on which there is normally room only for illustration of the product itself, perhaps with a serving suggestion or its immediate referent. Rarely can the printed advertisement or pack do what the television advert can do and illustrate the product in active use.

Printed advertisements exist in two forms, as a shape, usually a rectangle, by itself on a page or as a shape on a page in competition with often competing advertisements. In either case there must be a mechanism within the printed advertisement that arrests the scanning eye, when its gestalt properly says (or does not say) "Stop and look at me" to the reader. This is achieved by the nature of the picture itself, or by the presence of something interesting—perhaps a brand name or a joke. When this has been achieved the reader can consider the advertisement as a rectangle in isolation from the remainder of the magazine. The color contrasts within the advert, both typographic and picture symbols, can then be addressed.

When the advertisement exists as a shape on a page with competing adverts the successful advertiser must question the mechanisms used to make it stand out from its competitors. If it does not stand out, it will not be seen at all. The second question concerns whether the advert contains the information intended customers need in a form in which they can understand it and act on it.

When looking through the paper we are probably scanning to see what catches our attention. When we are in store we may be doing this but the committed shopper will also be looking out for specific products. Both pack and advert rely on conspicuousness. The shopper needs to be able to see the pack in competition with others. In the store blatant conspicuousness of the pack itself may be required, particularly for the case of an unfamiliar product. In the advert, any mechanism, even one totally unrelated to the product, may be used. For example, although mystery may be used to arrest the eye at a particular position on a page, mystery is the last thing required of the pack. There may be exceptions to the latter for products designed to be purchased on impulse.

We shop with reference to object recognition. Categorisation of visual stimuli as objects is a learned ability and we recognize objects through knowledge. We recognize meat and vegetables in the raw and hence in transparent wraps. This implies that, for speedy recognition, pack designs ought not to contain objects unfamiliar to the shopper. Knowledge is based on pictures and words, and product recognition, particularly of unfamiliar foods, depends on the legibility of both. The speed with which this can be achieved is only limited by the reader's visual acuity and scanning speed.

Within specifically defined areas of the store manufacturers compete to have their product type and label recognized. Ease of search is increased with familiarity of pack and with difference between search target and background noise. The more different the target from the background the shorter is the search time. At one time the British dairy counter appeared beautifully homogenous with uniform palish whitish shelves. It now contains ugly, still light, but fiercely yellow colors of butter substitutes and the varied saturated fruit colors of chilled yoghurt and dessert packs—the one time recognition by pack shape has been made more difficult by the addition of color. Adults prefer to select by shape, but children do this by color.

We all pick out and recognize colors as a whole; it does not depend on being able to recognize its elements of, say, hue, chroma, and lightness. Specialist skill is required to separate these individual components of color. In verbal communication there are very few colors upon which most of the population agree. The color memory of the average person is poor because there has been no evolutionary pressure for it to be good. We have evolved to assess colors in combination, hence we are highly sensitive to differences between colors when they are placed side by side. Therefore, conspicuousness of objects within an environment may be enhanced or depressed by using appropriate color contrasts based on our physiological response characteristics (see Chapter 2).

Uses of advertising color within the food industry are considered with respect to packaging and newsprint. For the latter the different print qualities available in magazine and local newspaper publishing are taken into account.

Colors have functional uses, for example, ripe tomatoes are colored red.
Color is used as a referent system, a color in the design refers to the color of the product being advertised.
Colors are used in contrast, including those in symbolic combinations, such as those in national flags or denoting brand.
Colors are used to reinforce the flavor of the food product contained within the pack.
Colors are used for product differentiation as well as product copying
Colors are used to deliver cultural messages.

FIG. 5.3. Uses of color in advertising and packaging.

USES OF COLOR

Color adds to the cost of printing, therefore there must be positive driving forces for its use. Color has many specific uses within advertising and packaging (Fig. 5.3). Each use is discussed more fully.

Functional Colors

Colors are used to make pictures more realistic. Tomatoes are colored red and grass is colored green. The display of a red tomato is used to infer that the product that is the subject of the advertisement or pack is made with fresh ripe tomatoes. Whether this is true or not we may not be informed.

Color is used as a code. For example, red is used to reinforce a hot scene, blue for one that is cold; red is used to depict femininity, blue for masculinity; red denotes arterial blood, blue, veined blood. Obviously the use of *red* or *blue* is not sufficient in itself, the context must be made clear before these specific connections can be made in the mind of the reader. So, advertisements of rum that comes from the sunny West Indies may contain yellow to bring tropical warmth to a beach scene and to the advertised product.

Individual colors have become identified with product ranges. Examples include white and pale colors for dairy products; bold, saturated colors with product pictures for breakfast cereals; and dark, rich colors for tea and coffee. All these work through contrast, which is brought about by areas of adjacent colors. Pack colors often follow the color of the contents. Examples for fruit juices, desserts, and sweets include: pale green for apple, green for lime, pale yellow for pineapple, and deep purple for blackcurrant. Associations have also arisen from other sources. Baby foods tend to be marketed in the pastel shades normally associated with infant clothing.

Color is used to illustrate traditional values, perhaps to associate pictures of the green countryside with healthy products of the green countryside such as butter. Similarly, the value of chocolates is equated with the value of gold. In the proper context black is used to imply sophistication, as well as cost, in such products as chocolates and liqueur.

Referent System Colors

Often in advertisements colors of items or scenes in the body of the picture are the same, or the same color character, as the color of the product itself. Color thus is used as a referent system for the product.

A common use relates the colored picture of the source of the product with the product itself. Thus, strawberry jam usually has a picture of a red strawberry together with a red stripe on the jar label.

Color is used to inject a mood into a picture. For example, one that features happiness, sexiness, warmth, or homeliness is designed to create an association of these fortunate feelings with the product. The designer hopes that the advertisement featuring the drinking of beer in the warm-looking gently colored homely pub will induce such feelings in the potential customer that will carry through to the purchase of the beer itself. Similar associations occur in colder climates when the warm colors of the beach are used to publicise the ice cream and in hotter parts of the world when cooler colors of the pool are thus used.

The tendency for children to respond to colors is exploited when advertisements for products designed to appeal to children contain fun colorful characters. For some children the character has overtaken the product. Instead of asking for a particular dessert they ask for a "Bob the builder." More subtly colored works of artistic pictures are designed to be noticed by adults, who tend to respond to shapes. Thus, brashly colored confectionery is designed to appeal to children while the beautiful shapes of the uniformly colored chocolates carefully placed in a shaped, compartmented box appeal to their elders.

Although the pack does not have room to picture flights of fancy, there is sometimes room to show serving suggestions. The high technology of the nondescript looking boil-in-the-bag white fish steak is shown in the home situation connecting it with a real meal. Inclusion of peas or green beans on the plate with the sauce-covered fish gives the designer opportunity to use a range of greens to liven up the bland looking fish. Natural products are not of one single color but contain a number that are closely related. This treatment adds a welcome darker contrast to the pictures of many fish products.

In advertisements there is usually a good color match of the referent object with the product picture. This is essential as both are seen at the same time. Similarly, the color on the label of the jam jar ought to match the jam that can be seen through the glass of the jar. However, a good match may not be necessary when the picture is of something inside the pack. For example, colors of breakfast cereal products in the picture on the pack frequently exhibit no color match with the product itself. In this case the brighter colors on the pack are designed to attract the shopper's eye. Such mismatches are generally disregarded or not noticed by the customer. However, in general we seem to have learned to be more subtle about observing the particular shades of browns of our preferred brands of chocolate and brewed coffee or tea. If the picture of the product is to match the product under all lighting conditions then the colors need to be a non metameric match. This is not always possible. For example, such a color picture of bloomed fresh beef cannot be obtained because there are no permanent colorants that match the peaky nature of the meat's spectral reflectance curve. Hence, under some illumination conditions the color of the picture could adopt an unfortunate mismatch with that of the beef.

Colors in Combination

Although many hues are used in pack design, the main source of contrast is based on lightness difference. Many hues may be used either as darker or lighter elements of a design. Colors that appear to be suitable for one product may not be suitable for all. For example, in the United Kingdom, although yellows are used for the lighter or darker elements on cereal packs, it seems to be unsuitable for teas, for which pale red is often used. White is still occasionally used as the lighter element in some older tea brands, but this probably dates from times when multicolored printing was more costly. Dark browns and blacks are commonly used for coffee and tea but not for dairy products, since they recall neither milk nor the countryside.

Specific colors and color combinations are strong identifiers of brand, for example, Heineken green, the red and white of Campbell's, the Kellogg red, black, and white, the Birds Eye dark blue, red, and white, Batchelor's red and white, and the red and white or silver of Coca-Cola. Products are also color coded; the cyan of Heinz beans and the purple of Cadbury's Dairy milk chocolate are examples. Close control of color is essential to continuing brand success. The color sometimes becomes the company and Cadbury specialist outlets heavily feature their purple. Colors are used as specific flavor codes. The red that goes with the white and silver of the KitKat chocolate bar wrap is combined with green or orange for the special mint- or orange-flavored versions. Here color is being used as a direct referent code. However, the traditional distinguishing quality image of the silver KitKat wrap has been discontinued. Changes in brand design are normally made slowly as sudden alterations may lead to a drop in sales. This occurred with Babycham when the bambi and distinctive bottle shape were abandoned in favor of a blue bottle with a bold yellow 'B.' The original design was rapidly restored. Label colors may be registered as part of the company brand image. For example in the UK, Cadbury has exclusive rights to its trademark color purple (first used in 1915) on confectionary packaging.

Colors that are too pale (lightish and grayish) suggest that the pack has been on the shelf too long and fading has started. Where poorer quality pack substrate is used, whites tend to be lightly tinted so that there will be no viewing condition under which the color will look dirty. Colors too dark and grayish indicate dirt from excessive handling.

Colors always exist in combination and there are specific rules of contrast existing within the design world. Successive contrast, for example, occurs as we flick through the pages of the magazine or scan the store shelves. Cones in the retina are then in a state of continual nonadaptation.

Simultaneous contrast occurs when colors are adjacent to one another. When complementary colors are adjacent they tend to be exaggerated. Such effects are exploited by Heinz with their can label and beans content. When the can is opened the product color looks a richer color. Cadbury also uses this effect with their purple wrap and milk chocolate. Packs of Birds Eye and Weetabix products contain adjacent complementary blues and yellows and in the butcher's shop fresh beef is displayed with green herbs or plastic leaves.

Undesirable effects may arise because of simultaneous contrast, in which a color is affected by the color of its surround (Giles, 1985). For example:

Dark hues on a dark ground, which is not complementary, appear weaker than when on one that is complementary.

Light colors on a light ground, which is not complementary, appear weaker than on one that is complementary.

A bright color against a dull color of the same hue will further deaden the dull color.

When a bright color is used against a dull color, the contrast is stronger when the latter is complementary.

Light colors on a light ground, which is not complementary, can be strengthened if bounded by narrow bands of black or complementary colors.

Dark colors on dark grounds, which are not complementary, can be strengthened if similarly bounded by white or light colors.

Although adjacent complementary colors tend to reinforce, adjacent noncomplementary colors tend to weaken, but their strength can be regained if a thin white or dark line separates the colors.

Contrast of extension is another visual effect occurring when colors are combined. This is an effect of color patch size. To achieve a balance, the areas occupied by some colors may need to be different. For example, a saturated yellow should occupy only one quarter of the area of its complimentary violet. Smaller areas of color tend to be relatively more vivid.

Spatial relationships occur in color combinations. For example, on a black background, light colors seem to advance while dark colors recede. On a white background the reverse tends to occur. Warmer colors and purer colors also tend to advance. Other contrast effects among the tricks of the designers' trade include contour and vibration contrasts, optical mixing and balancing of differently weighted colors. Part or all of the pack can be constructed with transparent or translucent materials. However, it is the designer who combines colors to create an effect of pseudo translucency.

In the real world of the market place and as our eyes scan the shelves, colors are modified both through simultaneous and successive contrast. Vision aspects of constancy, contrast, and harmony are discussed in Chapter 2.

Manufacturers may change pack designs from time to time. This is because our liking for color in packages and advertisements changes, just as our liking for the use of color and design in other parts of the environment changes. Changes are accommodated in the words, typeface, pictures, and colors used on the pack. Nevertheless, the manufacturer needs a stable logo that can be fixed into the minds of consumers. Yellow has always dominated Colman's mustard packs, normally with red and black for dark contrast. The impact of the Cerebos salt pack has been improved mainly by changing the dark blue, light blue, and white combination to dark blue, mid-blue, and white. The Typhoo tea pack was given a more striking shelf image for the UK marketing boom of the 1960s and 1970s. The color emphasis was changed from mainly gray with red, black, and white as contrasts, to mainly red with black, white, and green for contrast. In these cases, the manufacturers largely managed to retain their traditional logo colors while significantly increasing their on-shelf impact.

Colors to Reinforce Flavors

Field-dependent people permit factors of the environment to influence their perceptions of specific sensory stimuli. For example, for such subjects, sensations of flavor or sweetness of fruit juices can be increased by intensifying the fruit color. This effect may well have an evolutionary basis, as fruits possessing a greater intensity of fruit color tend to be riper, sweeter, and more strongly flavored.

In a similar way, preferences for toffee flavor differ with the color of the wrapper. This arises as a part of our learned "social DNA." For example, many people in Britain use wrapper color and shape to recognize flavors and textures of Cadbury's Quality Street confectionery. Sight of a familiar wrapper increases expectations of flavor and texture. For this reason wrapper colors of chocolate products were rarely changed. Now, however, such associations are being eroded as manufacturers strive to modernise brand image.

An apparent misuse of belief about color occurred during the Pepsi 1996 relaunch. The company widely and strongly advertised that the only change made was to color the can blue. This was because the color was postulated to possess a modern, cool, exciting and dynamic image. The message appeared to come over to the customer that the new can color would intensify perceptions of product flavor and coolness. The lack of immediate success was probably due to the fact that pack color works subconsciously; it does not respond to the fanfare of trumpets—see the Britvic example below. We have a resistance to being told something that is patently not true, but this need not hold for longer-term propaganda.

Color in Differentiation

Colors are used to make one product pack different from another. For example, in the soft drinks market Britvic, who now successfully use black to differentiate their product, ignored the reds of competing cola drinks. The contrast between Britvic's quiet success can be compared with the failure of the Pepsi strident relaunch (above).

Conversely, colors are used to assist in making one product the copy of another. This has happened with many products including soft drinks and breakfast products. Many store own brands picture color combinations that are very near to those traditionally associated with established brands.

Intense competition in the alcohol marketplace has livened up the drinks area of the store through differentiation often using bright colors and more sophisticated designs. Some labels were designed to appeal to underage drinkers. Certainly in the early days some of these alcoholic lemonades incorporated comic figures into the design. The most blatant of these figures have disappeared, but colors are still bright, vivid, and unethically designed to appeal to the underage drinker.

Cultural Messages Delivered by Color and Appearance

Product total appearance induces expectations. The images produced by bottles of rosé wine seen on store shelves in North American stores are influenced by at least six variables (Moskowitz, 1985). These are bottle size (large/small), type of cap (screw/cork), wine origin (France/USA), bottle color (green/brown),

TABLE 5.1. Examples of Cultural Taboos of Marketing Significance

Society	Color	Association
Iran	Blue	Funeral color
Egypt, Syria	Green	Nationalist color
Japan	White	Color of mourning
Nicaragua	Brown, gray	Disapproved
Latin America	Purple	Regarded as despicable
Malaysia	Green	Jungle sickness
Finland	Blue-green	Russian occupation

neck wrap (no/yes), and label type (simple/ornate). The image of perceived quality of wine was increased by the presence of a cork, a neck wrap, a brown bottle, and a US origin. The image of a dry flavor was increased by the presence of a cork and a French origin. It is possible to use this approach also to develop package designs for new products.

There are hidden cultural messages occurring within packs. For example, a man does not give his girlfriend chocolates because she looks hungry. Specific examples of design taboo occur in the international world. Many countries falling within the marketing areas of multinational companies have color and appearance taboos (Table 5.1). Similarly there are colors of celebration. In Europe many manufacturers advertise their products using colors associated with their particular national flag. Common examples are the green, red, and white for Italy, and the red, white, and blue for Britain, USA, and France. Marketing in Britain the French use French language words on their label to ensure the connection is made with their country. Colors can be characteristic of specific days of national celebration. Examples are red and white for St. George's Day, and red and green for Christmas in the United Kingdom, and green beer for St. Patrick's Day in the USA. In Japan the colors of celebration are red and white. At times of personal celebration and calendar events such as Girls' Day and New Year red rice, red fish, and traditional red and white sweets and cake are eaten.

TYPOGRAPHY

Advertisements are created for a specific community having specific inherited and learned characteristics. Hence, interpretation is culturally as well as personally dependent. We read best when we are familiar with the language, typeface, and page set-up. Legibility of words, colors, and pictures is needed to get the message across, but legibility may be downgraded in favor of the creation of a specific atmosphere in an attempt to make the display more memorable. Particular forms of typography induce images of different historical periods coupled with an individual character. This visual fashion is important to the way we accept the message.

Legibility is affected by paper quality, press registration, lay down quality, and ink density. Within such constraints print legibility is determined by the size and character of the letters, the spaces between them, the words they constitute, and the distance between successive lines. Larger type does not necessarily increase legibility. This is because the larger the type, the smaller the amount of reading falling within the normal eye span and the larger the number of eye movements and fixation pauses necessary. Spaces are also important, spaces between letters and a clear space after the full stop greatly improve the visibility of the sentence. Reading comfort is aided by suitable length of line, for example, ten to twelve words per line is the limit for roman type. Use of serifed letters and an aesthetic print quality increases word recognition. Speed and comprehension change significantly with typeface. Times New Roman is one of the better faces from this point of view. Some typefaces affect the ability with which older persons and those with vision defects can read. Ten, eleven, and twelve point are found most frequently. At normal reading distance the smallest size is optimal for younger readers, the largest for older. These findings do not apply at larger distances, say, on the bottom shelf of the store when many of us are obliged to kneel on the floor to decipher prices (Chapter 2).

COLOR REPRODUCIBILITY AND EFFICACY IN PRINTING

Newspaper printing quality varies considerably. In Britain this ranges from the higher quality of the weekend magazine to the lower quality substrates of the local weekly free papers. The most obvious difference between these extremes is the range of color possible. Greater ranges of hue and lightness are possible with glossy substrates. Lower quality paper absorbs ink and the coarser fibre structure increases diffuse light scattering. This results in colors of lower ranges of chroma and lightness. For example, there is a direct relationship between paper gloss and the density of blackness that can be achieved. Blacks on a glossy substrates possess L^* values of between 10 and 20, while on coarser papers they are much less black having L^* values as high as 40. (L^* is an expression of the measured intensity of perceived lightness, where an intense black has an L^* value of zero and intense whites in the order of 100.) Similarly, for saturated colors there can be color differences of 30 ΔE^* units, glossy papers having the capacity to yield brighter, higher chroma colors. (The ΔE^* unit is a measure of color difference, where one unit is approximately equivalent to a just noticeable difference.)

One of the more severe areas of competition in the local advertising press occurs within the restaurant and take-away food businesses. Reproducibility of repeat advertisements may be very poor, colors differing by perhaps as much as 10 or 20 ΔE^* units. In the higher quality daily press this color difference normally is reduced to a more acceptable 2 to 4 ΔE^* units.

The quality of definition achievable on pages of some magazines is so good that high quality artwork featuring comparatively small differences in color can be attempted. There can be problems of perhaps hue shift with week-to-week reproducibility but this is not normally noticed.

High picture quality cannot be expected from the poorer substrates used in the cheaper free press. Black and white line drawings are usually successful, but small pictures of pizza and pasta are reduced to an indecipherable mess. What is meant to be a tempting mouth-watering illustration becomes an unrecognisable jumble of colors. Printing definition and substrate quality are too poor in such publications because the range of colors possible is comparatively low.

Problems can still arise with better quality newsprint if they feature large blocks of color. These effectively reduce the range of other colors possible and this reduces color differences, contrasts, and discrimination within the advertisement. In other words, the message carried is made less decipherable. On the poorer paper often used by the local press the use of color is reduced to that of differentiating one fast food outlet from a competitor. That is, color use is reduced to one of contrast.

The matt coarser finish of poorer quality substrates supports neither lightness differences nor large color ranges. This low quality of reproduction means that the color itself is less important; it certainly cannot be produced with any degree of accuracy. What is important is that one advertisement displays different colors from another. The situation is very different with higher-cost gloss substrates that support the "works of art" of the modem sophisticated advertisement.

The translucent nature of newspaper means that print shows through the reverse side. Show through color differences of 3.5 ΔE^* units are very noticeable, particularly with large color blocks.

COLOR CODING

Color coding helps the warehouse person, store shelf-stacker, as well as the shopper. Dependence upon the product color coding is more apparent in packs than in adverts. Each flavor of potato crisp, for example, has its own color. Walkers Snack Foods potato crisps (chips) add a bright colorful impact to the store with their color-coded products. Once there was a logical connection between color and flavor but the number of flavors has increased and product differentiation is now the aim. Ready salted crisps are red, cheese and onion are blue, salt and vinegar are dark green, barbecue are black, prawn cocktail are pink, smoky bacon are deep red, beef and onion are brown, cheese and chive are yellow and green, and roast chicken are orange, Worcester sauce are purple, tomato ketchup are dark red, pickled onion are light green, and spring onion are bright green.

Cheap or 'value' products possess their own color coding, both Tesco and Sainsbury stores using a code consisting of dark blue and white stripes. In this way colors mean what the store or the manufacturer says they mean. Deliberate color coding not necessarily directly associated with the product is also used. This helps shoppers to identify desired flavors quickly. Heinz (UK) uses clear color coding to help harassed or inexperienced mothers. Labels on savory products for four month old babies are pale blue and white, for those who are seven months a mid red and white, and for one-year-olds red and dark blue. Labels on dessert products for babies of four months are green and white, and those for one-year-olds green and blue.

Colors come to mean just what advertisers want them to mean in terms of brand markers. Nescafe instant coffee powder, launched in 1939, is brown, it yields a brown product, and so brown is an appropriate color for label and bottle cap. Blend 37 marketed in 1955 was given a green top. The launching in 1965 of freeze-dried coffee represented a significant flavor improvement, so a golden jar-cap was chosen. The production of Gold Blend decaffeinated coffee in 1978 posed a problem. Green symbolising safety or good health would have been an ideal color. As this was already in use another easily identifiable contrast color was required. Blue and purple may then still have been too dangerous for food products in the UK (because in the 1950s they were associated with medicines) and orange is used for tea and cocoa. Yellow, once reserved for custard and mustard, is now used stridently with cereals and dairy products. Hence, the main color available was red. In the early days of decaffeinated coffee many consumers thought that the taste was not as strong as the caffeinated equivalent. As red communicates strength, it became the brand color for all Nescafe decaffeinated coffee.

Different brands of cola drinks are the subject of intense competition. When aggressive red was established firmly with Coca-Cola and Pepsi, Britvic were successful when they coded soft drinks black. In the days when such practices were allowed, Coke made it a condition of selling that they 'owned' the color red in that particular section of the store.

ANALYSIS OF APPEARANCE

Commercial studies of impact appearance are normally made using consumer focus discussion groups. For food a more formal analysis of appearance has been developed. This alternative but complementary way of looking at a scene is described in Chapter 1. Analysis is made using three types of vocabulary—expectations, visual criticism, and physics measurement. This type of approach can be applied to advertisement and package design.

Total Appearance comprises

Visual structure, that is, sizes and shapes of the pack/advert design and of elements within it.

Figure 5.5

WARM

COOL

HARD

IMAGE SCALE PAT. 1106334
(株)日本カラーデザイン研究所

Now, space is taken up with better design and creating perhaps mystery to attract the reader's eye to the advertisement.

Certain 'rules' occur as part of pack design. Measurements taken from a selection of food packs sold in stores are shown in Fig. 5.4. Areas of the pack occupied by brand name, product picture, and product name were measured. The percentage occupied by each is shown plotted against total pack area.

The familiar brand name (♦) occupies the smallest percentage of the total surface area; it increases gradually with pack size. Pictures (▲) normally only occur in packs greater than an area of approximately 20 sq in. This is because small pictures are indecipherable. As already described in this chapter this rule is sometimes broken to no good effect in advertisements on local papers. For larger packs the picture size is related to pack area; the bigger the pack, the bigger the picture.

Regarding the area occupied by product name (■), small packs can have a small or large part of the area occupied by the name. However, the longer the name the smaller the percentage area occupied because of the inherent length to width ratio of readable script. That is, the name can only occupy a large proportion of the area if it is small, such as 'eggs,' or if the pack shape area is distorted. An example is the cheese product 'Dairylea,' a name written across the diameter of a circular pack. Similarly, at the large pack end, the normal product name such as 'Cornflakes' cannot occupy a large proportion of the pack because it is a long name and for legibility the length/width relationships of letters must be preserved. An over large letter size cannot be read easily because of visual scanning problems—see Typography above. Hence there is a restriction on the relative area occupied by product name. With imagination some of these limitations are overcome. For example, on one current pack the area occupied by 'Parmigiana regiale' (a very long name) is abnormally large because it is written several times across the pack.

Kobayashi's Color Image Scale outlined in Chapter 2 (Fig. 2.1) can be used for plotting color combinations. Some of the word images for single colors are indicated in Chapter 2. Such a plot showing food packs available in 1986 in Japan is shown in Fig. 5.5 (Kobayashi, 1998).

Although cool hard colors were once considered unsuitable for foods, color combinations from all parts of the diagram are now found. Such diagrams are useful when exploring and considering where and how a product is positioned with respect to competing logo and pack designs.

These examples illustrate the fascinating almost natural laws that govern our visually perceived constructed world.

6

Expectations, Color and Appearance of Building Façades

EXPECTATIONS

The presence of a public house is normally announced with a freestanding signboard. This acts as a major impact focus. Such signs are colorful in their own right and the sign painter is able to use considerable skill and wit in its commission.

The façade or external furniture provide the impact colors that announce to the potential customer the presence of the restaurant or store. Is there any pattern to the colors used, or, do they just occur at the whim of the proprietor? It has been suggested that the selection of suitable colors (in this case for office buildings) depends on the location, but color desirability is based on the two psychological factors of *attractiveness and familiarity* (Sato *et al.*, 1997).

As might be expected there are in fact driving forces governing the choice of many of the color contrasts declaring the site of the hostelry. Greater competition tends to govern the link between high-impact color contrasts and cheaper venues. Higher restaurant prices are expected from venues with more discretely colored façades. Color arouses the senses. We often welcome a certain amount of arousal but high-impact colors tend not to be preferred—their sole use on cheaper food façades is to arouse. The sophisticated eating venue is more likely to be *attractive* than the low cost take-away establishment striving for survival among intense competition. Both types of venue, however, may be *familiar*.

The presence of large stores may be heralded with name and logo colors from the nearest motorway. Smaller stores in less affluent areas may be placed unobtrusively within the housing area they serve. In better-off areas they may be sited in more urban surroundings and many of the rules governing choice of impact colors are similar to those that apply to restaurants. Colors used on small shop façades tend to be dominated by blinds protecting windows from the sun; greens and browns are widespread. There is a tendency for the former to be used for grocery, meat, and fruit and vegetable shops, while brown may be more apparent for bakeries, but no firm rules apply.

Following the scheme outlined in Chapter 1, expectations arise from sight of the façade of an eating or drinking venue (Fig. 6.1).

Included in Fig. 6.1 are specific identification expectations associated with façades. These relate to whether we are comfortable with the type or brand of business advertised by the façade and whether it is convenient for us at that particular time to sample products we expect to be served on the premises.

Colors used in building façades have been noted in towns and cities around the world. Colors of façades have been summarized using the Color Zone Diagram created by designer Paul Green-Armytage and described in Chapter 2 (see Fig. 2.2). On this diagram can be plotted the *character* of colors used in each city and town area.

DRIVING FORCES FOR RESTAURANT IMPACT COLORS

Driving forces governing choice of colors used in restaurant and other façades are geology, the heritage industry, brand sponsorship, dominant population type, tradition, and ethnicity. Towns and cities used to illustrate the driving forces include Bedford (England), Mendoza (Argentina), Miami (USA), Old San Juan (Puerto Rico), Palermo (Italy), and Stockholm (Sweden). At each of the sites the exterior colors of a random selection of restaurants, cafes, and bars were noted during daylight in whatever weather conditions existed at the time. Thus colors recorded were as they existed, whether brand new, faded by the sun, or covered in dirt, as the potential customer looking for a meal might have seen them. Comments below refer to eating venues except where stated otherwise. Color categories refer to Fig. 2.2 and are noted in square brackets [ ].

Visually assessed safety: Have I been safe on previous visits to venues having this type of façade?

Visual identification: Does this façade advertise the type of venue I require, perhaps store or pub?

    Visually assessed recognition of the façade logo and associations with respect to a known brand, cuisine, or brewer?

    Visually assessed convenience: Is it convenient to partake of the advertised cuisine or brew at a price I wish to pay?

Visually assessed usefulness: Does this façade advertise the service I require?

Visually assessed pleasantness: Does this façade look good, have previous visits to venues with similar façades resulted in pleasant experiences?

Visually assessed satisfaction: Does this façade transfer a feeling of future satisfaction?

FIG. 6.1. Expectations arising from façades.

Geology

In places where a great proportion of the building is constructed from exposed local stone, it is this that provides one part of the façade color contrast. An example is Stockholm, Sweden, where the natural stone is gray, often subtly colored. Here one-third of observed venues possess stone as one of the significant contrast colors [*mid-grayish*]. Reddish, brownish, and greenish [*strong, deep* and *dull dark*] are the most popular dark contrasts to stone, while paleish, yellowish, and whitish colors [*pale*] are the most common light contrasts to stone. Of nonstone façades reddish, greenish-bluish, blackish, and brownish colors [*strong, deep* and *dull dark*] were the most popular dark contrasts, while paleish, whitish, and brownish [*pale*] were the most commonly occurring light contrasts.

The Heritage Industry

The cities of Old San Juan, Puerto Rico, and the Art Deco district of Miami provide contrasting examples of cities where what has been interpreted as past color glories are being recreated. The clean color quality tends to be the same in both areas but Art Deco colors tend to contain more white than those that appear to have been used in the days of the Old Spanish Empire.

In Old San Juan, a total of 35 observations were made, 31 of which contained white as a light contrast [*whitish*]. The darkest contrasts were mainly, in order of use, greenish, brownish, grayish, yellowish, orangish, bluish, and reddish. The wide range of dark contrasts also included yellowish-brown, purplish, and pinkish hues [*bright, vivid, mid* and *strong* and *dark grayish*].

The total number of observations in the Art Deco district of Miami was 11, of which ten had paleish and yellowish light contrasts [*pale, light*]. In contrast with San Juan no white colors were observed. The most popular dark contrasts in Miami were bluish and greenish [*bright*].

Brand Colors

Out of a total of 61 observations of eating places in downtown Mendoza, Argentina, 17 used brand colors to supply impact. Of these, 11 were dominated by Pepsi (red, white, and blue), four by Coca-Cola (red and white), and two by MacDonald's (red, yellow, and white). The façade either followed the logo colors directly or the impact derived from logo-colored chairs and tables situated on the pavement outside the restaurant.

In addition to these three multinationals there are a number of well-known restaurant and public house chains vying for business on the high street. Most demand that their particular impact brand colors or logos are used [often *vivid*].

Population Type

Bedford, England, has a large population of college and university students. To cater for these, many low-cost restaurants and take-away businesses have opened in the town. The fierce competition has resulted in the use of high-impact saturated reds and yellows [*vivid*]. Variation is added by the presence of other saturated colors used to advertise ethnic food based companies [often *bright, vivid*]. Driving forces here are solely used to create differences and the need of the business to be noticed. Customers of higher price more sophisticated restaurants catering for the smaller population of those who can afford to patronize them appear not appreciate the use of strident façade colors. That is, by changing the color the identity of the restaurant is changed.

Tradition

Tradition is obviously a major driving force for façade colors. Two examples are given, one concerning building tradition, the other concerning design style.

At various stages in British history the Tudor style of building construction has been fashionable. This now consists of black and white, mostly vertical stripes derived from black-painted wooden support timbers and between them whitened mud or plaster fill. Concentrations can be found in cities such as Chester, England, but there are many other examples of this type of black and white dominated façade from different periods to be seen around the country [near *black*, near *white*].

Italy is famed for its beautiful design style, particularly of accessories, clothes, and cars. In many modern urban areas architectural design is undistinguished, but layout and design of restaurant signboards often displays an echo of a favorable industrial style. In Palermo, Sicily, there is no blatant screaming of competing façades, but gentle reminders of the presence of a place where the inner person can have their culinary desires satisfied [many are *mid*].

Folk Custom

In Mediterranean countries many façades are white, as were houses and many cottages in the British countryside. This color is easily obtained from chalk and limestone, from which it is traditionally prepared by mixing the powder with milk or size. It protects walls from the weather and from heat, it is hygienically and psychologically clean, it transforms the appearance, and reflects heat [near *white*]. In parts of Greece white inside and outside of houses was said to keep the plague away, even cracks between paving slabs were painted (Lancaster, 1996). In Britain, lime was used as a fire retardant and walls and thatch were painted. Here white lines painted around entrances to buildings were once designed to keep witches away.

There is a tradition that the frequent use of red and yellow paint on Scandinavian houses was in imitation of the brick and stone of the grand houses. In Sweden, China, the USA, and Japan timber houses were traditionally painted red for protection against insects and rot.

Other Driving Forces

Façade color is also linked with ethnicity. Examples include the exotic colors used to create impact for Indian or Chinese restaurants [often *bright* or *vivid*]. Sometimes façade colors are linked more with function, for example, brown façades for bakers, green for greengrocers. These are the most commonly used colors used for window blinds [perhaps *deep*, *mid*, or *dull*].

DRIVING FORCES FOR STORE IMPACT COLORS

Impact colors used on exteriors of large stores and store chains provide a contrast with those used to advertise restaurants. Colors of a random sample of 38 UK stores were noted. Of these, 37 used two impact colors, one used three. The colors are listed in order of use in Table 6.1. The most used color contrasts were white with red, green, or blue; the frequencies are listed in Table 6.2.

A number of conclusions can be drawn from this small study:

Impact colors on smaller stores are high-impact brand colors [usually *vivid* or *bright*].

Impact colors on supermarkets may also be brand colors. However such stores are large; hence there is no visual competition and no need to use

TABLE 6.1. Impact Colors
Used in UK Store
Façades

Color	% Stores
Whitish	68
Reddish	37
Greenish	32
Orangish	16
Bluish	13
Yellowish	13
Blackish	5

TABLE 6.2. Impact Contrast Used
in UK Store Façades

Contrast	% Stores
Whitish and reddish	26
Whitish and greenish	21
Whitish and bluish	8

high impacts. Morrisons is an exception, always displaying the strident yellow and black contrast logo.

The greatest contrasts observed were saturated red plus white, and yellow plus black. These high contrasts were generally associated with stores that have or had more downmarket origins.

Comparatively lower contrasts are found on stores that have a history of being more upmarket. Examples of these are orange on brick orange, and white on a pale brick red. This conclusion of the link between sophistication and more gentle contrasts is similar to one of the driving forces that exist for restaurants, as noted above.

There are probably far fewer individual driving forces for store colors. In the traditional British town environment there are far fewer food stores than restaurants and the former tend to be larger and serve more customers. Hence, linking store impact colors with driving forces other than price may be irrelevant.

REFERENCES

Green-Armytage, P., 2001, Colour zones, explanatory diagram, color names, and modifying adjectives, *Proceedings of the International Color Congress*, Rochester, eds Robert Chang and Allan Rodrigues, 2002, SPIE—The International Society for Optical Engineering, Washington, 976–979.

Hutchings, J. B., 2001, Colour contrasts in advertising—façade colours of food and drink consumption venues, *Proceedings of the International Color Association*, Rochester, NY, eds Robert Chung and Allan Rodriguez 2002. SPIE—The International Society for Optical Engineering, Washington, 72–75.

Lancaster, M., 1996, *Colourscape*, Academy Editions, London.

Sato, M., Nakayama, K., Natori, K., 1997, Selection of suitable colors for office buildings based on their locations, in *AIC Color 97*, volume 2, *Proceedings of the 8th Congress of the International Colour Association*, Kyoto, Tokyo, The Color Science Association of Japan, 859–862.

7

Expectations, Color and Appearance in the Store

EXPECTATIONS

Expectations of the store are multi-faceted because it is where we might do our shopping, perhaps our courting, perhaps it is a place where we shelter from the rain. It is the cheapest form of nonhome entertainment and a community center, the weekly venue for coachloads of pensioners. The day can be spent anonymously browsing the shelves without fear of being pestered by assistants. It is a meeting place that is largely comfortable and safe. We can watch other customers, decide who is being self-indulgent, who eats alone, who is eating a balanced diet, who has a dog or budgerigar, and who is the frenzied shopper not stopping to check prices and carrying their basket to minimize checkout time. However, in-store seats are a scarce resource. Often the only place where we can sit down is the cafeteria where the customer is encouraged to spend more money.

Store windows play a role in the seduction and creation of expectation. There is no aesthetic point to completely covered windows as the store remains invisible, there is no customer communication, and nothing to tell the passer-by what is for sale. Only in those pubs habituated by criminal elements in London's East End are curtains kept closed. They, in contrast with the store, do not want casual visitors.

Whatever the reasons for visiting the store, our expectations are similar to those we have when visiting any other space (Fig. 7.1).

Shopping for many equals stress. It ought then to be the management's aim to provide a comfortable and satisfying environment. The store employs a visual merchandiser who makes it easy for us to find what we want—at least that is the theory. This individual is also responsible for making the place clean, tidy, warm, and a store to which we shall become so tied that we will always make our purchases there. They seek to tie us in with loyalty cards, pledges that there will never be more than one person in front of us in the checkout queue; they may provide crèches, self-scanning systems, money-off vouchers, and offer help with

Visually assessed safety: Will I be mentally and physically safe in here?
Visual identification: Is this store appropriate for what I want to buy?
 Visually assessed recognition: Do I associate particular irritations or concern factors
  with this type of store or brand?
 Visually assessed convenience: e.g., Does the store look busy, is the car park full, have I
  the time to shop?
Visually assessed usefulness: e.g., Will I find what I need?
Visually assessed pleasantness: Will my visit be an enjoyable one, or how can I minimise
 shopping stress?
Visually assessed satisfaction: e.g., Will I be satisfied when I have completed my visit to
 the store?

FIG. 7.1. Expectations on visiting a store.

the packing. If you get tired of actually visiting the store the computer-owning customer has the option of shopping on line. At first, it was the function of the store only to sell goods; nowadays the prime function is to sell the store. The store strives to love us and to give us confidence. To this end 'mystery shoppers,' 35,000 in the USA, are employed to check on store performance.

Quality, like beauty, lies in the eye of the beholder, or, in this case, the eye of the customer. Customer satisfaction can be measured as the perceived quality balanced against needs and expectations. The store measures changes in policy, whether this is layout or management style, against answers to three questions. These are, what pleasure does the customer feel when visiting the store, how aroused or excited does the customer get in the store, and what is the customer's willingness to buy?

Once we are inside our success as customers in responding to the environment depends on a number of factors:

How well we interact with the store as a brand, that is, will we go there? If we need only a few items we might go to the more conveniently situated and personal small shop. Alternatively, the big store may be nearer on our mental map of stores and car parks.

How well we react within the store, that is, will we explore the store?

How well we react with other customers and staff, that is, will we tolerate the presence of others, will we ask for assistance?

What general store image do we perceive, that is, do we regard the store as, perhaps, selling high-quality merchandise, or does it have a service-orientated strategy? If the former, the perceived high-quality merchandise must be accompanied by features that give an image of a high-quality store environment, if the business is heavily service orientated there must be an increased number of staff willing to be helpful in view.

How well we feel emotionally in the store, that is, do we find the store arousing or sleepy, exciting or gloomy, pleasant or unpleasant, or relaxing or distressing?

Similarly, the store's success resulting from these customer responses also depends on a number of factors:

How much money did we spend, did we spend more than we planned?

How well did we react to store and staff, do we have a negative image in the eyes of other customers?

Did we communicate our concerns, desires, and complaints to staff? This is important because if no one communicates staff have no opportunity to do anything about customer complaints.

Will the customer return for more?

What have large serve-yourself stores done for the customer? For fresh produce they have increased the number of varieties of fruit and vegetable, fetching exotic fruits from half-way across the globe. These certainly have injected a profusion of color into the average meal. On the other hand, self-service has had its consequences. Careful selection by the shopper has driven plant breeding towards sacrificing flavor in favor of uniformity of appearance. Thus produce may have been subjected to 12 or more separate applications of poisonous spray before reaching the store. Stores have brought an increased range of packaged foods to the shoppers' attention but their greatest success has been in influencing the improvement of wine quality worldwide.

The remainder of this chapter is concerned with a history of store appearance, store appearance design, and aspects of the store affecting the customer in person.

EVOLUTION OF STORE APPEARANCE

In the mid-nineteenth century the street market was the source of supply for the great mass of the population. The greatest numbers of street sellers were to be found on a Saturday night when, along with nearby shops and under the illumination of gas lights, members of the working classes purchased their Sunday lunch. Foods of variable quality were bought cheaply and sold cheaply. Safe storage of perishable foodstuffs was not yet common and the provision of fresh foods among a rapidly growing industrial population was still not possible.

Technological advances contributed to the success of the store rather than the market as a place of food sale. In 1806, Nicholas Appert, a Frenchman, devised a system of bottling for the preservation of meat, vegetables, and fruit. Six years later, Bryan Donkin substituted tin for glass and founded a canning factory in the East End of London. In the 1830s American Jacob Perkins invented the ice-making machine and in 1876 the first refrigerated ship, the SS Frigorifique, sailed from Buenos Aires to Rouen with frozen meat. Thomas Lipton opened his first shop in Glasgow in 1871 and by the turn of the century the Maypole and Home and Colonial stores were each operating more than 150 shops in the United Kingdom. In the towns, multiples continued increasing, largely at the expense of the market trader.

However, wherever there is a community there is probably a store. Around the world, stores are in different stages of development but the pattern of evolution is similar. For example, although the first store selling only packaged goods opened in New York in 1907, in the U.K. this did not occur until after the Second World War. Approximate dates included in the following historical sketch apply to Britain.

Until the late 1940s stores were places of personal service. Foods were often held in bulk and customers queued to be served goods individually weighed or counted to order into paper bags or wrappers. Kerosene, used for lighting, was poured and served at the same counter. There were few competing brands except for products such as biscuits and confectionery. Retail prices were probably not displayed; the goods were marked in code with the price paid by the grocer. The shopkeeper may have had a multi-tier price system, the amount charged dependent on the customer's appearance of wealth, and whether the goods would be exchanged for cash or time would be needed to pay. The premises were unhygienic, dark, dusty, and smelly, but the store played a vital role within the local community. There were always suspicions of the quality of more expensive foods. Pepper was sometimes bulked out with shop dust, ginger diluted with flour, ground coffee mixed with burned grains or pulses, and whisky diluted with tobacco juice and water. Hence, the market was ready for branded packaged goods of reliable quality.

Variation in store appearance was large. In contrast to such sombre interiors were the stores of J.J. Sainsbury. In the 1880s he specified the use of brightly colored decorated wall tiles, mosaic floors, and marble counter tops to make it easier for his staff to clean. His stores, by the current standards, showed a great advance in shopfitting and cleanness, and they were lit by gas. They were in town, well within a row of other shops but never on corners, which were seen as being sites for banks. In contrast, the normal grocery shop was not so hygienic, sacks of flour and sugar, for example, being open to contamination.

The first Tesco shops opened in the early 1930s, operating on a philosophy of pile it high and sell it cheap. Jack Cohen brought the principle of the street market to the shop. Inside was a small counter but fronting the store on the pavement were two pyramids of goods on open display. The customer could see what was on offer before going inside. Outside was a 'frontsman' selling from the display. The shop had roller shutters rather than glass windows.

Shortage of goods and staff during the Second World War led to the opening of Britain's first self-service store by the London Co-operative Society at

Romford in Essex in 1942. Since then, self-service has grown to dominate the retail goods sector. All items became pre-packaged and accessible and there was no member of staff behind a counter to get in the way of sales. Priced packages became available for examination and could be rejected without embarrassment when too costly. The shopper did not need to take a list of clear choices to the grocer and packs served as a reminder of what was needed as well as what was not needed.

Cohen opened his first supermarket in Britain in 1956. The floors were laid in red Marley tiles and Tesco red appeared on the ticket rows. Red and white blinds were placed above the provision counters and red lights were used to make the meat look better. Their aggressive visual style of the 1970s with Day-Glo window posters and trading stamps contributed to a cheap image. In 1977 the new livery of red on white was adopted for national advertising and in-house displays. The management aim was for simplicity, quality, and cleanness of style and an upgraded store appearance replaced what had become rundown and shabby.

In general, during the 1980s the pendulum had swung too far into self-service. Pre-packaging had lowered the quality image of some fresh foods so counter assistants were employed to serve in areas such as delicatessen, bakery, butchery, and fish. Areas displaying fresh foods became relatively small, self-contained, more easily controlled, and could be easily and regularly cleaned. Employees staffing these counters represent the store management and are responsible for projecting the company image. Employees are instructed not to sell and not to judge; their role is one of help and advice.

The physical design of the store evolved in the USA. The supermarket precursor was the 1920s grocery–meat–produce store that appeared in California and Texas. This was the time and the place in which cars first became widely available, paved roads multiplied, people started living further apart, and refrigerators were becoming a standard fixture in the kitchen. These factors, with the advent of packaging, combined to produce appropriate conditions for the flourishing large store. In 1916 the first Piggly Wiggly store was designed as a maze through which shoppers were guided past all the shelves until they reached the till at the far end. This design proved too customer unfriendly and shelving was set out in lines, thus freeing the shopper to wander at will. Store size increased and customers got lost, but from the management point of view this was no bad thing as it brought more people in contact with more of the items on sale.

In the early days, customers were given wire or wicker baskets for carrying their purchases. In 1919, tired shoppers were relieved of their burdens and provided with baskets that were guided by tracks set into the floor. This proved too inflexible and the modern, lightweight, steerable trolley evolved with the invention in 1937 of the nestable cart equipped with child seat. In an attempt to reduce the 600,000 trolleys that disappear in the United Kingdom each year, wheel jamming mechanisms are being incorporated to prevent their removal from the store car parking area. The tiresome unloading/reloading process at the till will become obsolete when automatic billing of the trolley contents becomes available.

So, store appearance changed from the dark, comparatively dirty, smelly place in which counter staff served the customer by measuring out required quantities of foods into wrapping or greaseproof paper, through the totally impersonal experience of packages on shelves, to the present combination of self-service of sterile but colorful packs and personal service at clean specialist counters, all in a well-lit airy clean-looking environment. Live staff provides the management with the opportunity of giving the customer that added personal touch. However, each staffed section provides another opportunity for delay through the necessity of having repeatedly to queue, a sign of bad service. Thus the appearance image of a relatively convenient and painless shopping experience can rapidly change for the worse.

STORE DESIGN

Layout

The aims of a selling establishment are to attract the attention of potential customers, to ensure that they enter the store, to make them feel happy while they are inside, and to ensure that they see goods and services on offer to the best advantage. Design, color, and lighting play major roles in achieving these aims. Shop and supermarket design and organisation depend on what image is to be conveyed as well as to how it evolved. The image of the mall is mixed, some regarding it as a convenient all-under-one-roof experience, others that they are prisons easy to get into but claustrophobic and a long way from the exit. Labor costs associated with the low prices of the stack it high and sell it cheap image are obviously less than those for stores seeking a customer service image. The store, however, is not merely a geometrically defined space; it is a cognitive whole arousing its own expectations in each customer. It is a space in which we should all be welcome whether we are a brand habit die-hard, a struggling green idealist or a self-indulgent epicure.

To the customer, stores have distinct general images that occupy three dimensions in a perceptual space. The dimensions are the quality of goods, their price, and the variety of items stocked. On top of this we select the store for our social, personal, and our own transport and convenience reasons. We individually can aim to predict and understand our social behavior in terms of reasoned action. This is derived from our attitude towards performing the behavior and our belief as to what others think about our performing the behavior. Social norms contribute towards our "good" and "bad" behavior. For example, would we like others to see us entering a store renowned for its cheapness and low quality? In other words, using a store, as well as buying a product, depends on what we think and what others think of us.

Stimulus from the environment induces emotional states of pleasure, arousal, and expectation. The greater the pleasure derived from the environment the greater the temptation to linger therein and the longer we stay in the store. From these emotional states the shopper's actual purchase behavior can be predicted. Hence, the greater the trouble taken to achieve the appropriate store environment, the more likely it is that increased sales will follow. From each customer's viewpoint this is convenient because it forces the store to make it as easy as they see possible for the shopping chore to be completed. For those in a hurry popular items can be located near the checkout.

Multi-entrance stores built in a circle with aisles that radiate from the center are confusing and disorientating and most stores are built to a grid pattern. Aisles should not be less than four trolley-widths wide; those perceived to be long, say, greater than 60 to 70 feet, and narrow or crowded can easily be omitted from the shopper's tour. The wider the aisle the more the store design encourages the more relaxed atmosphere of the free-flow pattern of shopping, in which no one feels obliged to hurry. Where there is likely to be queuing or other hold up a much greater width is necessary. Normally we do not like crowding in stores, but it may be tolerated, perhaps grudgingly, at Christmas. This is unlike bars or cocktail parties where we may actively seek out crowds and expect to have to tolerate a high population density. Task-orientated shoppers attend to density cues more and perceive greater crowding than the more relaxed browser. Store layout significantly affects sales and customer-tracking studies are used to test layout efficiency in terms of drawing customers past those shelves and sales areas the management wishes to emphasise.

If products are not in view they will not be sold. Sales can be influenced by the height that products are placed on the shelf. The optimum shelf seems to be lower than eye level and there is a reduced potential for impulse sales for goods placed above or below. Store management wishing to emphasize an image of reliability and helpfulness ensure that all brands inhabit their regular shelf area. Here the customer can always find it. In some stores particular brands are moved around the store from time to time so that customers have to embark on a time-wasting search. The theory is that in doing so they might perhaps find a product not seen before and thus add to the goods purchased on impulse. This strategy involves risk as only approximately half the women who cannot find the brand they came into the store to buy will buy a substitute; the others will not buy anything.

The designer has an important job to do in making the large areas pleasant to experience. This is done through judicious use of color, decoration, and lighting. Colors appearing clean under one illumination condition may look dirty when viewed under store lighting. Hence, color selection ought to be made for the specific illumination used in store. This and good cold cabinet and shelf design makes merchandise appear more attractive. Good produce and product displays increase the customer's feeling of pleasurable expectation, but pleasure can quickly evaporate when the drab signs of wear and tear are evident. Store refurbishment is needed when the condition of display equipment appears out of date or when competitor activity demands it. Shoppers appear to approve as the process normally results in an immediate increase in profits.

Our expectations for bodily safety are comforted by the sight of fire extinguishers, emergency exit doors, floors free of merchandise, litter, and fluid spills, and properly lit exit signs which should be quietly visible from all parts of the store. Mental and bodily safety expectations are improved by the presence of customers similar to us.

Quality of lighting is vital to the way stores, produce, and we ourselves look. This subject is discussed in Chapter 3.

A Walk around the Store

Appearances change as we walk around the typical UK store. Prominently sited near the entrance is the manned counter at which the customer can interact with a member of staff and purchase small valuable items such as cigarettes. This area may be decorated in the company logo colors. Nearby are the high mark-up ready meals and fresh fruit and vegetables set against a background of black, perhaps mirrors, but more often in the United Kingdom, with green. Fruit and vegetables are situated near to the entrance because it creates an initial image of freshness, life, and natural color for a store filled with sterile plastic, paper, and card packaging. Where produce is brilliantly colored, the setting can be a neutral color—mostly black and white. This is the only department that can legitimately display dirt. However, we are so cleanness conscious that this dirt is probably cosmetic, perhaps clean potatoes boxed in dry peat. No dirt is permitted on most produce; all of it is cleaned, sometimes with surplus water adding to the weight paid for at the checkout. Even though fruit and vegetables have been cleaned, when we get it home it should still be re-washed to minimise spray residues. Overall we select using visual signals of quality; these are cleanness and neatness of appearance, brand name, a strong warranty, price, and image of the store.

Fresh produce may be followed by the dairy product displays. Whites and near whites of milk, cheese, and cream packs are probably stored together, now in an area unfortunately blighted by saturated yellow packs of butter substitutes.

Separated from this are the high mark-up multicolored fruit and sugar-containing yoghurt or milk dessert products, some aimed at children. The section displaying packed fresh meat, again, is a clean white, different types of meat sitting in color coded foam trays, perhaps with green to contrast with the red meat. The wet fish counter should be displayed beautifully with color contrasts provided by different species and translucency contrast provided by crushed ice. The bakery perhaps displays its quiet browns against pale colored shelves and black or silver colored wire trays.

Then, we move on to the glazed clean delicatessen counter, at which we may have to queue. Many clues to hygiene standards can be found here. We should see that dairy foods are kept apart from raw meats, which should be kept in a separate cabinet from cooked foods. Handling must be seen to be minimal during cutting and weighing, and separate implements should be used for cutting different foods. Premises and staff should be clean and all perishable foods must show an 'eat before' or 'best before' date.

Following the fresh produce displays and counters we move into the store and among the ranked shelving containing multicolored canned, boxed, and bottled products destined for the kitchen store cupboard. Carefully avoiding trolley contact, we come to the white horizontal and vertical freezer cabinets. Here we may be seduced by tubs of ice cream; according to the advertisers this is as good as sex. However, products buried in these cabinets have minimal visual impact. Frozen foods are hidden in opaque wrappings as we do not like to see a build-up of ice around the product. In contrast, composite products such as brightly colored pizza can be shown off to perfection in the chill cabinet. We might miss out the chill and freezer aisles altogether, as these areas can be uncomfortably cold.

High-selling items such as alcohol, water, and soft drinks are perhaps situated at the very end of the store, placed furthest away from the entrance so that customers will be drawn past lower sales volume goods such as household hardware items. Somewhere in store, perhaps close to an otherwise colorless area, are color-coded packs of potato crisps (chips) and snacks. On our way around the store we ought to beware of mowing down product pushers standing in the store aisle offering us a taste of something new. We are meant to feel a sense of obligation to purchase.

Signposting may be helped when the floor or wall color is used to identify department, perhaps whitish for dairy goods, brown for bakery, blue for fish, or green for fresh produce. These sections may provide the customer with more opportunities for dynamic interaction with members of staff and provide the management with opportunities to increase the pleasure of our visit.

Appearance factors contributing to the lowering of a store's image include pale, off-color, and frosty meat, bloody, sticky, dirty, and broken packs, dust and dirt on shelves or food containers, and dirty and untidy store employees, who are not polite or who behave in a visibly unhygienic manner.

Efficiency of the visual search for a pack in a complex color environment is controlled by the number of areas containing packs with similar color, the target pack size, and the complexity of background colors. This creates difficulty for the smaller store seeking to stock as wide a selection of goods as possible. The more items stocked the smaller the space allocated to each product. This increases search time and creates a dilemma, because if store size is increased the concept of a rapid shop is eliminated. A profusion of copycat products seeks to confuse further our visual search.

Whatever routine grocery and household items we require, they will be widely separated throughout the store. This is not an expression of Murphy's or sod's law; it is deliberate management policy to place milk, bread, eggs, and sugar far apart. High mark-up items will be prominently displayed; low mark-up items will be less visible. Cheaper own-brand goods are likely to be placed on the bottom shelves, the higher profit branded alternatives at eye level. Stores do their best to make it easy for us to spend money.

Merchandise should be left to sell itself, but some colors are more appropriate in food displays and eating situations. To be avoided in store are too many colors, colors that are too deep, modified colors such as yellow-green, and distracting colors. The other colors are the so-called appetite hues, discussed further in Chapter 8. On the meat counter white should be present to convey cleanliness; in other areas bright colors play the same role. Food the consumer sees as a result of lighting and color design should look good enough to eat. Some stores incorporate in their layout a profusion of near-saturated colors. Our visual comfort is decreased in such visually polluted environments.

In-store presentation for selling involves imparting a visual impression of cleanness, neatness, and symmetry. Even the stacking of shelves should be performed meticulously. Particular attention should be paid to the alignment of labels of bottled, tinned, and packet goods. Dented and damaged packs should be discarded. Customers are usually particular in not buying such packs anyway. On the other hand (unlike the advertising display), the arrangement must not look so perfect that the shopper is dissuaded from disturbing it. Product or theme presentations liven up windows as well as the inside of the store. They also make products stand out on the shelf. The principles of display are included in Chapter 9.

Arriving eventually at the checkout, we set our minds against the deliberately placed temptation of top-selling, brightly colored confectionery brands sited there and commence study of what other shoppers have in their trolleys. While daydreaming, we desperately hope that we have taken advantage of all the 'buy-one-get-one-free' offers to which we are entitled. If we have taken such advantage we must remember to look carefully at the checkout monitor to see that the computer has been told about the BOGOF offer. Finally we expect a dazzling smile from a hard-working, defenceless, penned-in cashier before we push our purchases out to the totally secure carpark. One day we will be able to use virtual reality to tour the virtual store selecting virtual items for our virtual trolley. This will be less painful than real shopping, but without the pleasure some customers derive from human contact and the physical act of walking the aisles.

CUSTOMERS

Sensory Marketing

Sensory marketing is a psychophysical ploy designed to provide sensory landscapes comprising light, color, smell, sound, and touch stimuli to accompany and to enhance favorable perceptions of the goods on sale. The customer is deemed to need such extra stimuli, which are designed to distract the shopper from the boredom of serried ranks of modular shelving bathed in intense white light.

Although there are no hard and fast rules for the music that accompanies television advertisements, it is believed that classical music is more effective than Brit Pop or easy listening music in increasing customer time and spending in store. Selling may also increase when the product is placed within its original environment. For example, playing French music may increase sale of French wines. We dance to the tune of the store as the speed of the music dictates the speed at which we move around. A slow tempo leads to slower progress and increased spending. If the volume is low the sound can be ignored; if not, people are driven away. A 20-minute shop accompanied by tinny renderings of Christmas carols is sufficient to irritate most shoppers. All day exposure must lead greatly to staff stress at the busiest time of the year. As if unasked for noise is not bad enough, in some places we are now forced to tolerate in-store radio.

Added smells in the store are common. Among the benefits of the in-store bakery are the delicious smells channelled to other areas of the store. Smells come in bottles, so the luxury of a bakery on site is not essential. Blatant circulation of bakery smells has generally been abandoned because, like the story of the Pepsi blue failure, such stimuli only work if the customer is generally unaware of their presence (Chapter 5). Plans are under way to isolate areas of related produce in its own terrain, adding appropriate smells and sounds. Asda has tried out smelly till receipts at Christmas, while Superdrug filled its stores with the smells of chocolate on Valentine's Day.

Tactile behavior, say touching of hands, is equated with friendship and warmth and gains attention. This form of communication often increases our feelings of interpersonal involvement, in turn increasing our feeling of commitment and liking of the employee and store. Thinking laterally, store managements are now considering the introduction of a range of 'feelie' packs that reproduce the feel of the product inside the wrap.

Thus the customer is given a more complete sensory experience in the absence of the proper stimulus. This makes a change from the scents of sterile plastic wrappers and fridges. Its aim, of course, is to tempt us to think of buying products that we would otherwise not normally give a second look. Certainly, increasing the pleasantness of stores tempts us to stay, and where we are tempted to stay we are more likely to be tempted to spend. Such tactics may increase spending by some individuals, but will many otherwise loyal customers be alienated?

In-Store Irritations

Pack labeling can be a source of great irritation. The Co-operative Group in the UK has identified the following "seven deadly sins" of dishonest food labeling (Anon). Some are worse than others, but they can be categorized under the following headings.

The Illusion. Labels which do not tell the whole truth on the front of the pack on the grounds that customers would not want to know. For example, products called *Mince and Onion – minced meats and onion gravy* in which the main ingredient is mechanically separated chicken.

Weasel Words. Common in the store are misused and sometimes mean-ingless words such as *traditional, home-made, original, farmhouse, natural,* and *fresh*. *Natural, pure,* and *fresh* applied to packaged and processed foods mean nothing. *Farmhouse* has a specific meaning for loaves, long and split down the center. Used with any other food it is also meaningless. *Fresh* is often applied to fish that has been frozen, trans-ported, and thawed. *Home-made* is often applied to factory-made goods. The Food Standards Authority recommends its use is restricted to goods made in the home in a "domestic setting." *Original* and *traditional* are used to give false authenticity when a newer version appears on the market. The *traditional* product should have existed for a considerable period of time. *Original* can mean the opposite—a new version entirely (FSA, 2001). Included is the use of terms like *fish steaks* to describe products cut from large blocks of fish.

Rose Tinted Spectacles. This category includes pack designs, photographs and words which give a misleading impression of the product inside. Examples include the use of small plates to make products look bigger or by retouching photographs.

The Bluff. Some labels make the product seem special because of something left out or added, when actually this is normal practice. For example, dry pasta where labels claim the product is *free from preservatives* when in fact no dried pasta contains preservatives by law.

The Hidden Truth. This category includes labels containing important information that is difficult to see. For example, the legal name of the product is often printed in small letters sometimes on the back of the pack. For example, the 'with sweeteners' statement which must accompany the legal name is sometimes hidden where the minimum percentage meat content is usually found also.

The Half Truth. Labels that inform the customer on the front of the pack what is not in the product instead of what is. For example, products that make claims such as *80% fat free* which means "With 20% fat", a not negligible amount. Alternatively, the freely given information that a product is *95% fat free* is not accompanied by the even more helpful information that it contains 40% sugar.

The Small Print. Labels, for example, in which nutritional and ingredient information are printed in small type in colors that do not contrast with the background.

Our perceived mis-classification and mis-shelving of products is a great source of irritation. We search first for what we think is the appropriate department of the store. Some of these entry levels, such as *bakery* or *dairy*, are clearly labelled and defined. Others, such as *home baking* or *foreign foods*, are not. Fresh beef is found under meat, but meat products such as corned beef will be found under *tinned goods*. Grapes are *fruit*, but dried grapes such as sultanas may be classified with flour in *home baking*. Likewise, pudding rice is filed not with rice but with custard powder in *home baking*. Nuts may be divided between *snacks* and *cooking aids*, each without shelf reference to the other. Cornflour is probably with *cooking aids*, not with flour. A mis-classified product may be regarded as missing or out of stock; this enhances neither the customer's temper nor the store image.

Shelves can be too high to reach in comfort or safety. They can also be too close together. A glass jar, perhaps containing preserves, that is jammed tightly into the shelf may be left where it is, especially if its removal is likely to bring others crashing to the floor.

Regular customers will find their way around their store quite happily and where there is a problem the helpful signposts high in the ceiling will probably be a most useful guide. This is so until large notices are hung over the aisles and signposts are removed from view. Large splodges of color (orange is very effective) catch our eye with irrelevant messages and obscure what would have been valuable direction signs. Products such as financial services have no separate shelf space so stores tend to become littered with large unfortunately-placed hangings. Such practices are particularly frustrating at Christmas, that is, from the beginning of September until the end of January—approaching one-third of the calendar year. Then, during the busiest period of the year, aisles are narrowed and products are moved around the store as the management attempts to pack in more and more food and seasonal products. Hence, during a period of love and kindness, the conditions provided are ideal for trolley jams, customer rage, frustration, and anger. One in five of us have experienced shop rage.

Shoppers make a number of navigational errors (Titus and Everett 1996).

Orientation and destination identification errors: getting lost, misidentifying product or store sections

Destination selection errors: mistaking where particular products can be found.

Environmental assessment errors: being in a section and dismissing it as a likely location for the product that ot does contain. For example, being in the pasta section and ignoring it as a potential location for a limited selection of Italian wines.

Product analysis errors: not knowing what a product is or what it should look like.

Product recognition errors: perhaps mistaking one product for another

Some of us are grazers, removing products from the shelf as we move around the store in an orderly fashion. Stores are designed to meet the needs of such customers. However, some are planned shoppers who have not got much time to get their shopping done. These people rush around disturbing the quiet progress of the grazer, who is more likely to spend extra on impulse purchases. Larger stores can stock at the entrance area premium items, such as sandwiches and cold drinks, required by buyers for lunch on the move. This is good—if there is a till to relieve these shoppers of their money.

Queuing is a form of stable equilibrium relying on the voluntary cooperation of strangers who will probably never see each other again. It is a sign of civilisation, not practised by other animals, when we copy the behavior of the majority. Some customers may regard time spent in queues as periods for reflection on the problems of the world. To most it is a frustrating imposition on their time and being forced to queue twice or even three times in the same store is not viewed civilly. Even the sight of a queue is enough for many customers not to wait at the delicatessen counter. It is in the interest of the store to have sufficient staff manning such areas. Not only are they usually selling more expensive produce but also members of staff are the management's personal contact with the customer. It is difficult for a single employee serving a long queue to keep a smile on his or her face. However, the not-more-than-one-customer-in-front policy at the checkout is good, but only if there are sufficient tills and staff in store to cope with the rush.

Store marketing of fresh fruit and vegetables has resulted in a bland uniformity of product. Adding to this, European Community legislation has resulted in the reduction of a number of produce varieties. Within their own varietal classification, all tomatoes, for example, are the same size and shape to within hardly noticeable margins. The uniformity is not surprising; customers go to great effort to sort carefully through each individual fruit before selection is made. This pickiness has directly led to multi-spraying of herbicides and fungicides onto crops, ensuring that there is not the slightest trace of insect and other damage. If you have to pay high prices, and supermarkets are regarded as charging highly for fresh food, then you might as well select what you regard as the best. Flavor seems largely to have been bred out and signs saying "Grown for Flavor" admit to this. Similarly, dietary scares have led to fat being bred out of beef and pork animals. Unfortunately it is the fat, particularly the marbling, that gives rise to the best flavors. It is perhaps fortunate that the independent restaurant owner can still shop around for high quality produce; this helps us to remember the good old days.

Three aspects of pricing cause annoyance. Uniformity of pricing of different varieties of the same produce is arriving very slowly in Britain. It is beyond most of us to compare price per kilogram with price per pound with price per item. For those buying on a budget it is important that we do compare such methods of pricing, especially as one of the alternatives may have been air freighted huge distances at great expense.

When diligent customers reach home they will check their bill against the shelf price remembered or noted down during their progress through the store. It gives great satisfaction to reclaim money incorrectly charged by the store but it ought not to be too difficult to match goods on the shelf with the barcode and with the price in the computer. The protocol of shelf-stacking must be linkable with shelf tag and with the price charged at the checkout.

Over a long period customers have learned that ready-made meals are of lower quality and more expensive than those prepared at home. This has led to an expectation that it is inevitable that such meals are of lower quality. This in turn has led to stores continuing to sell meals of lower quality even though manufacturing technology has greatly improved and is capable of producing ready-meals of good quality. Some of the better meals purchased look as though they have been purchased in toto. Clever hosts will add items to enhance the home-cooked appearance to food they serve to guests. Such items include mushrooms and fresh herbs such as lemon parsley, basil, and dill. Similarly, own brand goods, such as beer, are of high quality and cheaper, but these may be regarded as not being good enough to give to guests.

When a product line runs out and that part of the shelf empties it is tempting for store employees to push other products along so that the gap is filled. The old price tag is not removed because it is used to inform stackers that replacements are needed. In this way, customers are charged the wrong amount.

Other shelving sins, particularly in smaller stores, include the 'special offer' tag that is positioned so that it is seen to apply to another product, one not on offer. Similarly, sometimes it is not clear whether the price tag on the shelf applies to the product above or below the tag. Spotting these shelf-labelling errors requires a sharp eye at the till and could leave the customer with a larger than expected bill.

Price tags seem designed not to be read. The normally sighted customer, when close enough, can read prices when labels are placed at eye level. However, not many of us can read the same label when it is placed sometimes more than a metre lower—especially when it has been tucked under an overlapping box of produce. Scrabbling about on our knees to find the price is not dignified behavior for a pensioner. Not all of us possess our full range of faculties or have retained a youthful capacity for energetic or spatial activities. Some of us are less mobile, and have deteriorating visual or hearing capacities. Of the population, 45% possesses one or more vision defects. Is there someone to help those in need of it? A help would be for the store to change layouts less often.

Shapes are more readily perceived when there are differences in lightness. The boundary between two adjacent areas possessing equal lightness will tend to disappear. In larger displays use of near equal lightness may not be important because other clues are usually available to accentuate boundaries. In small displays, however, such as shelf price tags, boundary visibility is critical, particularly to low visual acuity shoppers.

Not nearly enough thought has been given to the multitude possessing eye defects. Is there an outside door that opens automatically? Are labels readable? Are boxes left in the middle of the floor? Able-bodied people design able-bodied packs: are they legible, reachable, handleable, and openable? There are many elderly single people who need to cater only for themselves. Will they be able to open screw-top lids, plastics seals, can ring-pulls, shrink-wrap? Are portions small enough; are there patient assistants available?

Such concerns may seem inconsequential to the young and fit, but for those who are aged it is the way they have to live. On average it takes a senior shopper twice as long to complete their shopping. Fortunately, there are 'senior simulators' in Britain and certainly in Japan designed to reduce the young and fit to the less flexible, sensorialy deprived, more confined way of life of the old and those with mobility problems. Not only are there the obviously disabled but also there are the invisibly disabled people with asthma or hearing impediments. Simulators used widely in the retail trade will help the store adapt to the larger than ever proportion of elderly people in the community.

We might be less flexible because we are parents of young children. Many of the large stores provide rooms where babies can be changed. Some even provide crèches. Not all have carparking spaces reserved near to the store for mothers with small children and for those who find walking tiresome. Thought for the customer, able-bodied or otherwise, is slowly arriving.

Concerns of the Customer

Intellectual and emotional aspects of the visually derived appearance image raise concerns about the product and the way that it is produced. These concerns are focused on the manufacturer as well as the retailer. Sometimes it will be a matter of fact determined from the detailed list of ingredients, perhaps fat or sugar content. Sometimes it is a matter of informed opinion such as 'greenness' or 'healthiness.' A list of such customer concerns, and responses from the manufacturer and retailer, are listed in Fig. 7.2.

For most of the categories, manufacturer, retailer, and government fight to get the customer to believe information given about products and their manufacture and transport. It is impossible for the customer to verify most of the information given. However, some concerns can be verified using our own senses. Visual checks can be made for untoward growth of pathogens, cleanness of serving and eating areas, equipment, and handlers. Also, food that is meant to be hot ought not to be merely warm. Ingredients of products on the shelf can be checked for whatever the doctor says is bad for us. If we are bothered about value for money we can fight our way through the pricing maze. Hence, we can arrive at a common basis, say price per pound, for prices of different varieties of the same vegetable or fruit. The store ought to do that, but many do not.

Nutrition: nutrient content, medical fashion, such as roughage, cholesterol, are organic foods healthier?

Concern for environment: "greenness," genetically modified crops, pesticides and herbicides, conservation, packaging, organically grown produce, local produce.

Exploitation of children: i.e., targeting them with bright colors and designs, e.g., "alcopops," high-fat/sugar products such as blue margarine, bright green dressing.

Quality: including honesty of claims, misleading labeling.

Value for money: high prices, misleading shelf pricing.

Animal welfare: growth hormones, endangered species.

Lack of confidence: use of potentially misleading marques; see text p. 99.

Concern for the Third World: including fair pay, depletion of resources, unethical behavior of store chains.

Food risks: e.g., salmonella, unclean food handling and selling environments; additives and preservatives.

Food sourcing: food miles per product; see text p. 98.

Political and religious correctness: e.g., Turkish language on products for Greece, Arabic script on exports to Israel; neither will be widely accepted.

FIG. 7.2. Consumer issues involving appearance in its widest sense.

Many years have passed since downright poisons were included in food products, but relatively small groups of consumers cannot tolerate specific foods or ingredients. Sensitivity to nuts has resulted in in-store warnings of products likely to contain them. Those who are lactose- or sugar-intolerant generally know those foods to avoid, and if they do not, the label fine print gives such information. How-ever, many wine drinkers are intolerant of sulphite preservative. Although wine labels in the USA have to declare its presence, those in Europe do not.

Some foods travel large distances before arriving at the store. In Britain, sources of fruit and vegetables are marked on the shelf or product label. Some travel, however, is hidden. 'Greek yoghurt' does not travel only from Greece; it is made with milk from cows that live in Germany. Some stores have a policy of locally sourcing as much produce as possible. Many do not. Customer pressure for reduction of transport costs by stores first supporting producers local to the store has resulted in the Tesco policy of labeling 'local' all items produced in the UK.

As far as labeling is concerned (see Chapter 5), buying brands we trust and reading press statements regarding issues outside our direct experience will help us to shop more sensibly—if we wish to. Concern for food 'healthiness' is felt more by adult females than adult males. Both are more likely than children to choose foods that they themselves think are 'healthy.' Effects of hazards on the menu are more critical for children and the elderly.

The Co-operative Wholesale Society in Britain has addressed animal welfare concerns. They label 'intensively farmed eggs' as such and are working to reduce animal transportation time. They have a customer right-to-know policy that details their Responsible Retailing Charter, giving customers sufficient informa-tion to make informed choices. Other stores lag behind.

Of significant concern are seals of approval. There is a belief that a logo from an independent organisation on a food product means that it must have some intrinsic value over unendorsed ones. For example, the Family Heart Association, the National Osteoporosis Society, the British Dental Association, and the Football Association Premier League endorse different products. There is no indication on the sponsored products about what the logos mean. Normally they do not mean what we think they mean. For example, the Football Association Premier League endorses Lucozade. This is a sponsorship arrange-ment, and purchasers thinking that it is a quality mark or a health sign have been misled.

Standards also mislead. The British Farm Standard and British Meat Quality Standard logos may imply that meat bearing the logos comes from British-reared animals. This is not the case; these standards can be given to meat produced abroad. Information the customer wants is not supplied. The Lion Quality Mark used on eggs indicates only that they come from hens vaccinated against salmonella and does not indicate that the hens have been raised in any particular way.

All these concerns form part of the total appearance of the product.

REFERENCES

Anon, 1997, The Lie of the Label, a Report Calling for Honest Labeling, The Co-operative Group, Manchester.

Ajzen, I., and Fishbein, M., 1980, *Understanding Attitudes and Predicting Social Behavior*, Prentice Hall, New Jersey.

Birren, F., 1969, *Light, Color and Environment*, van Nostrand Reinhold, New York.

Boswell, J., 1969, *JS100 The Story of Sainsbury's*, J. Sainsbury, London.

Danger, E. P., 1987, *The Colour Handbook*, Gower Technical Press, Aldershot.

Donovan, R. J., Rossiter, J. R., Marcoolyn, G., and Nesdale, A., 1994, Store atmosphere and purchasing behavior, *J. Retailing*, **70**:283–294.

Doyle, P., and Fenwick, I., 1974, How store image affects shopping habits in grocery chains, *J. Retailing* **50**:39–52.

Engel, J. F., Blackwell, R. D., and Miniard, P. W., 1995, *Consumer Behavior*, Dryden Press, Fort Worth.

FSA, 2001, *FAC Review of the Use of the Terms, Fresh, Pure, Natural, etc., in Food Labelling 2001*. Food Standards Agency, London.

Gifford, R., 1997, *Environmental Psychology*, 2nd ed., Allyn and Bacon, Boston.

Heller, W., 1988, Tracking shoppers through the combination store, *Progressive Grocer* **Nov**:47–54.

Hine, T., 1997, *The Total Package*, Back Bay Books, Boston.

Hornik, J., 1992, Tactile stimulation and consumer response, *J. Consumer Research* **19:** 449–458.

Kaufman, C. F., 1995, Shop 'til you drop: tales from a physically challenged shopper, *J. Consumer Marketing* **12**:39–55.

Murrills, H. C., 1955, *The Displays of Canned, Packed and Bottled Goods*, Blandford, London.

Pegler, M. M., 1991, *Food Presentation and Display*, Retail Reporting Corporation, New York.

Powell, D., 1991, *Counter Revolution*, Grafton Books, London.

Russell, J. A., and Pratt, G., 1980, A description of the effective quality attributed to environments, *J. Personality and Social Psychology* **38**:311–322.

Scott, R., 1976, *The Female Consumer*, Associated Business Programmes, London.

Sonsino, S., 1990, *Packaging Design*, Thames and Hudson, London.

Titus, P. A., Everett, P. B., 1996, Consumer wayfinding tasks, strategies and errors: an exploratory field study, *Psychology and Marketing* **13**:265–290.

Wheeler, A., 1986, *Displays by Design*, Cornwall, London.

Worsfold, D., 1998, What's on the menu? An analysis of the hospital menu for food safety hazard, in: *Culinary Arts and Sciences II: Global and National Perspectives*, J. S. A. Edwards and D. Lee-Ross, Eds., Worshipful Company of Cooks, Bournemouth.

8

Expectations, Color and Appearance in the Food and Drink Consumption Environment

EXPECTATIONS

What is so special about a nonhome environment and what do we expect from it? Why do we want to eat or drink out; what are our expectations? The reason may focus on the home, such as taking the family out, wanting a rest from cooking, not wanting to drink alone, or to celebrate, to give the partner some peace, or to get away from the kids. Alternatively, reasons may focus on the venue, perhaps to meet old friends, play games, a liking of the pub atmosphere, to relax, to concentrate, or a liking for professionally prepared and served food. Expectations also arise from the anticipated eating and drinking in public in a currently fashionable and valued situation; fashionable, that is, to us and in the company of those whom we enjoy or at least tolerate. The external venue is a place of structured, mannered behavior, in which we act according to an established code of behavior; different codes for different venues. In the restaurant, for example, we experience a sense of well-being derived from a combination of self-presentation, the opinion of others, and the appearance of wealth; that is, wealth in terms of our being able to afford to pay for the current experience. Integral to the pleasures of consuming out is the ambience of the venue. The ambience, or decor, comprising furnishings, lighting, and layout provide the setting and summarize the mood we expect to enjoy. For example, comfortable chairs and rich colors encourage expectations of luxury, while brightly colored plastic furniture perhaps suits a more casual approach to life.

How can one venue possibly cater for such a wide range and variety of motives? With one exception, no venue is designed to cater to such a broad audience. Normally, of course, the hospitality business attempts to isolate and divide its potential customers into groups. Isolation may be by class, where entry depends on the amount of money the individual is willing to spend, or by regulation through a code of dress. Isolation may come through serving beer rather than the range of lagers enjoyed by the boisterous, monied, high-energy, noise-tolerant young. Those who are older will certainly resist the temptation of a long stay at such a venue. Isolation may be self-imposed by our predisposition as to ambience, food, and the size of the eventual bill.

Such barriers are lowered, for example, when all members of the community, irrespective of restaurant segment patronization, do duty by drinking a plastic cupful of luke warm tea in the wasp-laden atmosphere of the village fête. An exception is also provided by the large convenience chains, where, almost wherever in the world travellers find themselves, customers know what to expect.

In Britain there are many types of eating and drinking establishments existing within a distinct hierarchy, Table 8.1, of perceived quality and actual cost. To most British people each of these venues presents very clear expectations. This mental picture consists of sets of physical total appearance characteristics. At the expensive end of the range the picture will include staff in uniforms, chandeliers, lots of gloss, lots of transparency, and style. At the cheap end might be hard seats and soiled carpets, and perhaps torn upholstery. Each venue induces expectations of inner comfort, that is, type, quality, and cost of food and drinks, and outer comfort, that is, of intimacy and condition of lavatories. In general we seek an establishment that will supply acceptable goods and services for sale at a price appropriate to the environment and at a price we want to pay. Pubs and restaurants isolate, protect, and identify us (with the type of place it is); they also isolate, protect, and identify the owner (with the type of person he or she is, and with the type of customer attracted).

Three types of eating environment have been identified (Finkelstein, 1989). These are the *spectacular*, the *amusement*, and the *convenience* restaurant. Spectacular restaurants derive their spectacle either from the setting, such as the top of a tall building, or from the formal decor of sumptuous furnishings. Both places are expensive, and are where clients go to see and be seen. There are also two types of amusement restaurant, both of which can also be expensive. These are the theme restaurant, where food quality is of secondary consideration

TABLE 8.1. The Hierarchy of Eating and Drinking Places in the UK

Country		Town	
With Alcohol	Without Alcohol	With Alcohol	Without Alcohol
Restaurants	Transport cafes	Restaurants	Tea/coffee shops
Pubs		Bars	Sandwich bars
		Pubs	Greasy spoon cafes
		Working mens clubs	

to the decor, and the importance of dining restaurant, where the quality of the food is all. Convenience restaurants comprise the café, fast food, and sometimes ethnic eating places.

Although there is a wide range within each category, each type of restaurant possesses a characteristic appearance. That of the informal spectacular restaurant is focussed on the spectacle, perhaps featuring large windows at the top of the skyscraper. The formal spectacular restaurant has an impressive interior, perhaps complete with crystal chandeliers, tapestry-covered chairs, widely-spaced, beautifully-covered tables, marble fireplaces, wood-paneled walls and heavy carved furniture. In the reconstituted environment of the theme restaurant and encouraged by appropriate behavior from the serving staff the customer is expected to act out the reality. The room may be a fake, containing modern representations combined in a re-creation of a stereotypical image. There is a variety of importance of dining restaurants. There may be little attention paid to decor, but often items on view are carefully selected to be complementary in quality to that of the food. The fast food restaurant, on the other hand, is genuinely modern, tends to be brightly lit and have polymer-tile flooring, fixed, easily cleaned plastic chairs and tables, bare brick walls, and little soft furnishing to absorb noise. Around the walls there are likely to be pictures and, perhaps in the windows, models of the meals served. The ethnic restaurant may possess the full distinctive decor of the cuisine on offer, or possess a stereotype of the genuine decor, or merely display the token trinket or Chinese lantern.

The supplier of the food or drink is in business for the profit. The greater the cost of the meal to the customer, the greater the opportunity for profit. However, the restaurateur must convince the diner that the scenery of the restaurant, that is the decor, tabletop, manner of presentation, and service are equivalent to the cost demanded. Most of the factors comprising the scenery are appearance related. The restaurateur operates on the assumption that mood, a sense of relaxation or luxury, or intimacy exists. The ambience operates to further this aim and the staff operates to reinforce this mood by managing the emotions of the customer. The customer normally acts as a willing participant and normally must derive pleasure from a situation in which he or she is being exploited. Thus, appearance and ambience is less about food but a great deal to do with dining out.

Decor involves the creation of a scene from a space, a backdrop against which personal events can be played. All spaces, designed or otherwise, have physical properties that lead to five broad groups of expectations. When we enter a space, we make, possibly subconsciously, visual estimates of our expectations (Fig. 8.1). The extra identification expectations are similar to those resulting from viewing a façade. These concern my associations with spaces having similar appearance characteristics, and whether it is convenient for me to eat or drink there at the time of the visit. These expectations are dominated by the elements of appearance.

Visually assessed safety: e.g., will my physical and mental safety be secure, that is, will I
   be mugged by the clientele, poisoned by the food, or bullied by the staff?
Visual identification: e.g., does this pub serve meals?
   Visually assessed associations: e.g., will I be able to obtain a particular food in an
      environment of preferred cleanliness, comfort, privacy, and quality and price here?
   Visually assessed convenience: e.g., is it convenient for me to eat/drink here; how long
      will it take to be served?
Visually assessed usefulness: e.g., how useful will this space be, can I do what I want to in
   or with this space, will the staff be efficient?
Visually assessed pleasantness: e.g., how pleasant will be this space, food, and service, will
   I be happy and comfortable doing what I want to in this space, will the staff be friendly?
Visually assessed satisfaction: e.g., how satisfied will I be when I have concluded my
   business within this space?

FIG. 8.1. Expectations of food and drink spaces.

Our expectations will probably be satisfied when we go for a routine visit to
a venue we know well, one where we know we can get what we want in an
environment to our satisfaction and at a price we can afford. When this normal
venue is closed, we may be forced to eat food of an unaccustomed quality or
quantity from a menu offering different choices in a less desirable environment
and at a lower or higher cost. Hence, expectations of pleasantness and satisfaction
from our visit might be compromised.

To celebrate that special occasion, we might expect to pay more than normal,
possibly because the environment is more costly, and/or we expect a different
quality and/or variety of food, perhaps more sophisticated or home made. In such
a venue, physical safety might be greater but we might feel our mental safety is
compromised. The usefulness and pleasantness may be high, but the satisfaction
may not be as high because we know we are going to have to pay extra. At an
even more sophisticated level, perhaps when eating from gold plate in the
presence of the Queen of England satisfaction might be high, but will we feel
quite so comfortable as when we are drinking at our local pub?

The final example concerns the eating occasion in a geographical area in
which we are strangers. Our life experiences include the building of relationships
between the perceived external environment, the associated internal environment,
and associated food type and quality. When we are strangers in town this can lead
to our making errors in judgements of expected food or cost or environment. This
situation has been alleviated in some areas of the market by the presence of a
strong brand image offering product consistency. Associated with such venues are
clear expectations of imagined safety, identification, utility, pleasantness, and
satisfaction.

The physical environment plays a large part in how we feel when we eat and
drink out. From the images we receive we may feel comfortable or uncomfortable, at home or a total stranger, in a fashionable environment or one where we

do not wish to be seen, transported to a foreign environment or merely among unfamiliar shapes and colors, transported to an environment for which we feel sympathy, for example, the Irish pub, or transported to a modern environment of futuristic design, consisting of bright colors and glass. For many eaters and drinkers it is the environment that governs what they eat and the price they will pay for the experience. Success of these differing venues depends on images generated in the minds of members of the aimed for population segment.

When we seek outside entertainment, we have expectations based on our preconceived ideas about the event. If the actual service received is higher than our expectations, we have a positive service performance gap. This sets the standard for the next time we make a visit to the venue, when our expectations will be higher than those we had on our first visit. The aim of the provider of the service is to achieve a positive, or at least a zero service performance gap. This applies not only to the complete experience of the evening but also to each individual aspect, such as food quality, cleanness, or comfort. Disappointment with just one aspect of the event can spoil the whole experience.

This chapter looks at the evolution of the eating and drinking decor and features of the physical design. It also contains the results of a study of the characteristics and images of eating and drinking places, and summarizes color use in the internal environment.

EVOLUTION OF EATING DECOR

In twelfth-century England, refreshment could be had at religious houses and in cookshops and taverns in the medieval settlement. Appearance extras were minimal; the refectory decor was plain and constructed with stone, lath, and plaster, lime wash and wood. In the village the brewer's bush rather than the signboard advertised the availability of ale.

During the centuries that followed the fare on offer, as today, varied and evolved with the pocket and needs of the customer base. In the villages during the sixteenth, seventeenth, and eighteenth centuries, bread and cakes were available, sometimes toasted and steeped in ale, or perhaps given away with drinks. Because of fuel shortage the laborer may have taken his own meat to the local hostelry or bakery where it would be cooked for him. In the towns, more substantial and sophisticated fare was available. The appearance of the pub in terms of whole-someness can be gauged from the diary account of eating and sleeping in the eighteenth-century inn: "Nothing could have been nastier than our inn at Worksop; with ill cookery, stinking feather beds and a conceited fool of a landlady: but I endured, in my old quiet way, knowing that were much worse to be found." Similar conditions prevailed in London but not the entire eighteenth-

century city was filthy. Viscount Byng in his diaries longed for home territory
" . . . tourings and county visitings only serve to whet desire for London quiet and
London luxuries" (Byng Diaries, 1785–1794).

A late-eighteenth-century visitor to a butcher's shop near Green Park found
"the meat so fine and shops so deliciously clean; all the goods were spread on
snow-white cloths, and cloths of similar whiteness were stretched out behind the
large hunks of meat hanging up; no blood anywhere, no dirt; the shop walls and
doors were all spruce, balance and weights highly polished." And in a London
inn, "The cubicles were neat, the tables were laid with white cloths, and there
were delightful wicker-chairs to sit on" (Hope, 1990). Taverns, inns, coffee and
eating houses of London pre-date the legendary origins of the formal restaurant
generally attributed to cooks liberated when their masters went to the guillotine
during the French revolution.

In the mid-eighteenth-century Boulanger a Parisian tavern-keeper advertised
his food "Walk up, everyone who has a weak stomach. I will restore you." The
French for restore is restaurer, hence *restaurant* came to be used for places where
food was served. From the early nineteenth-century, cookshops, where ready-
cooked foods could be purchased, were in competition from coffee shops,
bakeries, cake shops, and cafes. At this time the public house survived to serve
the growing, more mobile population by offering more substantial food. The
interior appearance of the pub changed from the early single living room of
the laborer's cottage, to the provision of a number of small rooms (some
occasionally provided for women only) at the inn catering for locals as well as
travelers, to the nineteenth-century luncheon counter at which the customer could
stand and eat, to the present day larger, open, and more easily supervised space.

In the Middle Ages, hygiene standards were low. Although it was known that
infection was spread by contagion and airborne infection, there were open sewers
in the streets and marketplaces with animals free to roam among food waste. A
major concern of today's customers and public health authorities is cleanness of
serving, preparation, and eating areas as well as hygienic handling and storage
of foodstuffs. Above all, eating areas must be seen to be clean. Everything
accompanying a meal, that is, items comprising the tabletop, can be basic as long
as they project an image of cleanness. The mid-nineteenth-century saw an
improvement in the cleanness appearance and convenience of the kitchen when
gas began to replace coal as the preferred cooking fuel. In commercial kitchens
steam was also starting to be used for warming plates and turning the spit. Later,
the kitchen had the potential to became even cleaner with the development of the
electric cooker. Most household kitchens were equipped with gas or electric
cookers by the late 1930s. To be borne in mind when designing a home kitchen,
the reader of the Daily Mail Cookery Book of 1927 is advised "Let your text be:
Everything in the kitchen will have to be cleaned" (Hope, 1990). The modern
food critic comments on the apparent cleanness of the facilities offered, the

atmosphere, the quality of the food and drink, and the welcome received by the stranger. In contrast, health authority inspectors are concerned about the absolute cleanness.

A number of factors contributed towards the development of the rarely varying menu of the fast diner chains. These included the unreliable quality of food and the hidden expenses of a meal. Vegetables might not be included in the price and the bill might reveal the extra cost of a cover charge. In the case of the chain restaurant, what you see on the photograph is what you will get to eat at the price stated on the picture. In the USA, White Castle, founded in 1921, was one of the first branded restaurants guaranteeing constancy of product and service. On the outside they looked the same, white-walled with battlemented and turreted rooflines. Such brand design façades became readily detectable by the passing motorist ready to yield to the food impulse purchase. Later White Castle Diners looked the same on the inside also, with porcelain enamel panelled walls, easily cleaned and hygienic of appearance.

In London, the Savoy and later the Trocadero were restaurants for the wealthy. Joseph Lyons produced in 1904 his own version for the less well off. At the Popular Café in Piccadilly, diners could eat from a simple fixed menu or à la carte in pleasant Surroundings with orchestra playing. His clean, comfortable 1000-seater Corner Houses catering for London office workers followed later. Customers in Lyons' smaller teashops were served by 'nippies,' waitresses dressed in black and white and renowned for the speed and neatness of their service. Lyons attempted to bring the appearance of luxury to the lives of a greater proportion of the population (Hope, 1990).

The self-service store inspired the concept of the self-service cafeteria, introduced in 1895. Prepared food tempted hungry diners as they filed past. This is impulse buying applied to the restaurant.

The first plates consisted of square slices of bread, meat being placed on this trencher. These were replaced in the sixteenth century by the wooden trencher. Until the second half of the seventeenth century the knife, spoon, and fingers served as domestic cutlery, the fork, knife, and spoon were then brought together as an unmatched set. Individual bowls were introduced at the same time. Although Marco Polo introduced porcelain into Europe in the fourteenth century, its use did not become general until the early eighteenth century when commercial production began in Meissen, Germany.

PHYSICAL DESIGN

The design acts as a trigger for the diner's expectations. An expected design type will lead to more comfortable feelings than one that is not expected. Customer perceptions on entering a food and drink venue arise more or less in order. First perceived is the general ambience of colors, lighting, smells, temperatures, and sounds, which leads to an impression of whether there is room for your party, whether it is clean, comfortable, and of an appropriate standard and apparent cost. Second, when we get seated in the restaurant or have accepted the challenge of entering the pub, the quality of service on offer becomes evident. This includes the quality of staff and the furniture or tabletop. Third, after we have been served, we can judge the qualities of the products we have been given—appearance and expectations of the food itself is discussed in Chapter 9. Fourth, when we receive the bill, we can judge whether our initial judgement of visual satisfaction has been vindicated and whether the occasion has been value for money.

There are two major aspects in which the design of the eating environment is important. There is the initial impact that occurs when we look through the window or first walk into the room. If this engenders the wrong images, we will walk away without eating. Assuming this image fits our expectations, the initial image may assume less importance. However, to some customers the continuing image is important, for example, to the diner who finds it impossible to eat an Italian meal in nonItalian surroundings. Yet others, after the initial acceptance, become unaware of the environment, or, will only become aware if something is wrong. This degree of awareness of the environment fluctuates according, among other things, to our boredom level or to our state of involvement with other individuals.

Individuals change character according to the event because they play different roles at different times. For example, the same person may be the meek subordinate at a business meeting, the dominant father on a family outing, and the boring golfer at the sporting dinner. The same individual will behave differently in the same restaurant. At different events the organization of the ambience, meal, and service can change accordingly. Most of these occasions can be recognized by the color and appearance of the eating environment and casual observation of the occasion.

In the city it is the general view of the designer that it is the restaurant critic who must be won over. It seems that the inspiration of the environment greatly helps in judging the inspiration emanating from the kitchen. This does not necessarily apply to the general diner. Major critics visit places out of reach of most readers, both financially and geographically. Critics from the local paper tend to be more realistic about food and their reviews are of more relevance to most of those who eat out. Targets for designers change as society changes. For example, the past twenty or thirty years have seen the rise of the concept pub. Also raised have been such questions as: Is this place female or middle-aged or gay or singles friendly; is the carpark visually and physically secure? Managers who know of trusted taxi drivers and who have a list of unsafe areas in the locality tend to have an enlarged customer base. The same period has seen a mushrooming of ethnic cuisine restaurants. Appropriate ethnic total appearance through use of traditional colors and ornaments adds to the ambience of such venues.

However, the flair and intuition of the designer are on top of the considera-tion of physical layout. This book is not about the designer's flair but about the customer's perception of it. To a large extent, this is involved with the general ambience, the feeling, itself greatly influenced by the cleanness, comfort, and convenience of the design. Convenience includes such factors as whether the appropriate type of seating is on offer, whether the tables are too close, whether the tables are placed near doors or traffic lanes, whether there are attractive views from the windows, whether the space is flexible enough to cater for your party, and whether this is a venue having, or not having, facilities for children.

Constructions may be required to engender images of luxury and high cost, or the availability of cheap, fast service, or that of a special scene, such as the Australian outback, or whatever fashionable space expected to catch the eye of the high-spending, fashion conscious diner. From the customer's viewpoint there are two major elements to the scene. These relate first to the near view or to the tabletop. The second relates to the view of the ambience provided by the designer or the scene that is viewed through the window. For both elements there are a number of important aspects of design relating to creation of images of spatial dimension, cleanliness, quality, privacy, and comfort.

The total appearance of restaurant, cafe, pub, and bar consists of images derived from physical things. These comprise the visual structure, a series of surface textures, colors, glosses, translucencies, properties that change with lighting and time. The scenes are converted into images by the make-up and life experience of the perceiver of the physical scene (see Chapter 1). Design is a combination of inspiration and knowledge of materials, but there are rules. What physics aids are there to help the designer fulfil the brief requested by the commissioner of the design? Assuming physical criteria for spaces devoted to infrastructure, preparation, serving, and selling are met, what other customer-sensitive rules apply? These are discussed under the headings of spatial dimen-sion, cleanness, quality, comfort, and privacy.

Spatial Dimension

The most inspired design can be ruined if the physical layout of the restaurant is unsuitable. The body contact endured in the crowded pub is not appropriate for the leisurely, expensive dinner (Chapter 4). Form and function in the restaurant are satisfied in the main by considered placing of chairs, tables, and serving stations. This depends on the required flexibility, intimacy, luxury, and maximum numbers of diners to be catered for at one sitting. Layout of tables and chairs contributes to the comfort or discomfort of eating and serving and varies greatly with the type of establishment. Dimensions and distances associated with the dining process are recommended in space-planning standards (De Chiara *et al.*, 2001). In general the greater the cost of the meal, the greater are the dimensions and distances. Of course, guests with restricted mobility require more room.

Having outlined the mechanics of dining room organization, the designer can concentrate on creating and extending the general ambience. The decor involves the creation of a scene against which the personal events and the drama of the meal can be played. Scenes create images and the decor acts to control image formation within the individual. Restaurant decor is a created, sometimes blatant stage. Low lighting and rich red browns may create a stage for romance and intimacy, while high lighting and greenery may add to an ambience of health and activity. Furniture itself adds ambience. Perhaps tables that are a little lower than usual make the diner feel more secure and in control, while the size of the table adds to intimacy and a feeling of comfort. A soft surface to the table adds to sensations of ease and comfort. The many visually different interiors of restaurant in the high street are a compliment to the imagination of the designer and to the flexibility and science of the technologist. It is within the designer's skill and inspiration to create an ambience of, say, homeliness, smartness, femininity, and sophistication, in addition to the major images of cleanness, privacy, comfort, and warmth.

There are exceptions where we will tolerate and even enjoy a cramped meal. Epiphany lunch at Piana dell' Occhio trattoria, Palermo, Sicily, is a well-patronized family event with everyone dressed in Sunday best. Sitting on benches at tables 290 cm long, each customer has a space of 48 cm (that recommended for high-density banqueting is 76 cm). Benches are almost in contact, reducing the nonserving aisle width to zero. The distance between the ends of each wooden plastic-top covered bench table, the serving aisle, is such that it can only be negotiated by the thinnest of waiting staff. Once in place the customer is there for the full duration of the meal, at least two and a half hours. This meal defied all the laws of comfortable eating. Food was presented on plastic plates with plastic mugs for the water presented in plastic bottles and local wine served in jugs, the napkins thin and small but in plentiful supply. The decor is plain (5% chromaticness), but the emphasis is on cleanness of surroundings and quality of food. Such conditions would not be tolerated at commercial restaurants in Britain or the USA but, within the context of the occasion and the company, an experience not to be missed.

Form and function in the public house can be more complex than that for the restaurant. There are four functions performed within the bar room floorspace of the average British pub. These include a secure space behind the bar at which drinks are sold, perhaps a games area where darts or skittles is played, and two types of customer space in front of the bar. The last consists of the space used by customers standing or sitting near the bar as well as space in which there are tables and chairs used for those eating or engaging in more private conversation. In the well-designed pub these individual functions are blended to create a whole, coherent image. Color, texture, and lighting are used to isolate each function as well as blend them together. Customers are expected to make do with smaller personal dimensions. For sitting at the bar we are permitted a shoulder width of 40 to 45 cm. The recommended circulation width drops from 90 to 60 cm; the actual widths, of course, quickly drop to zero during busy periods. Some members of the population actually prefer it that way.

The different types of pub possess their own physical features, as identified by Kate Fox (1993) in *Pubwatching with Desmond Morris*. The serious traditional pub exists for the real ale it sells. It is plainly decorated, containing nothing contrived. There are no items of plastic, chrome, bright colors, or bright lights and no jukeboxes, games machines, or pool tables. The traditional country pub, now nearing extinction, may feature a games machine and have facilities for serving food. The town center version is less under threat from the brewers because it is popular with the wealthy young. The circuit or trendy or fun or theme or disco-bar pub possesses many of the standard features of the discotheque or nightclub, such as open-plan design, shiny surfaces, and bright lights, as well as areas of relative darkness and perhaps a small dance floor. There may be raised posing platforms for the more confident, extrovert clients. (The newest version of this is the 'style bar.' Doormen only admit those wearing specific high-cost designer clothes.) The family pub has large signs outside making it known they cater for children. Inside they look deceptively like 'normal' pubs, but may have the give-away clues of highchairs, brightly colored place mats, and smiling faces on menu boards. The estate pub normally caters exclusively for people living near rather than for a particular type of person. It acts as a social center for the local area. Often they are large, with more than one bar, perhaps a separate pool or games room, and perhaps a separate function room upstairs. It will have home-like features; the decoration will be comfortably shabby and have a lived-in feel. The student pub is a cross between an ordinary pub and a student-union bar. There may be a few traditional pub decorations, beer advertisements, and posters advertising student functions. Furnishings will be solid and well used design is unsophisticated with no attempts made towards color coordination or inclusion or the inclusion of stylish extras. The yuppie, stockbroker, and green-wellie pub can be found in urban and rural areas. It caters for urban professionals in the more fashionable area of town. They range from the classy wine bar to the upmarket pub. It will be attractively designed, fairly sophisticated and tasteful, more subtle than the circuit pub. It will probably not be themed and will have few home-like features. Hence, pub design, like restaurant design, is mainly about repackaging of old established ideas.

The spacious well-laid table is a joy to behold. On the other hand there is the airline meal. A number of factors erode the comfort of the diner. Among the irritations of flight are the hot foil dish, cutlery and condiments efficiently wrapped in difficult to open plastic, an unbalanced knife with inefficient cutting edge, hard meat, hot sauce, and nowhere to place a drink or the contents of the cutlery pack. Perceived physical safety is low and lack of personal space, poor atmosphere, smells, and time zone changes do not add to the joys of budget travel. However, there is plenty of variation of color, transparency, opacity, and gloss ingeniously packaged into a small space. In an attempt to deliver the appearance of a balanced meal, caterers include time-, temperature- and storage-sensitive greens. The sight of khaki or gray peas and beans is not appetizing to the British passenger. Fortunately, food pigments other than green chlorophylls are generally sufficiently stable for inclusion in chill-stored reheated meals (Chapter 9).

Cleanness

Cleanness is of concern to customers of eating and drinking establishments. Surveys commissioned by British brewers reveal that 69% of regular pub customers (Scottish and Newcastle Retail) put 'clean toilets with loo roll' at the top of their list of customer service requirements. This was seen to be more important than the quality of the ale or food. Les Routier hotel guests completing 'satisfaction' forms have one major concern, that of the cleanness and efficiency of the lavatories (see *Hotelkeeper and Caterer* 1 April 1999). At the Bah Humbug restaurant in Stratford-upon-Avon, concern about cleanness of the lavatories was emphasized by the presence of goldfish in the transparent cisterns. However, allegations of cruelty to the goldfish prompted their removal. Many diners work on the assumption that if the washrooms are dirty, food preparation routines will be dirty also.

Restaurant hygiene standards are of concern to the trading standard authorities; these can be classified as (Leach, 1996):

Cleanness of cutlery, equipment, crockery, glass, hands, and fingernails of staff, serving areas, the kitchen, chipped or cracked crockery.

Temperature: food at the correct temperature in salad and cold meat bars, hot food that is hot, food cooked properly right the way through unless ordered to be otherwise.

Food: salad materials well washed, food on display well presented, waste food to be in lidded bins.

General hygiene: no flies around the food, coughs and colds kept at a distance, staff hands washed after sneezing or blowing of the nose, staff fingers kept apart from the food, no soggy tea towels draped around the waist.

Customers of restaurants normally take standards of food hygiene and cleanness for granted; hence, expectations may already be high when entering the premises. A definition of quality involves the discrepancy between expectations and perception (Zeithaml *et al.*, 1990). In other words, where the venue fails to meet the standards of hygiene and cleanness expected, then customers will perceive it as offering a poor-quality service. The resulting emotional state is dissonance, that is, lack of harmony, and the customer will take steps to avoid the situation again. Hence, poor service will probably, in the United Kingdom anyway, not be reported, but the return business will be low. Customers are not all the same, however. The ultra-cautious will take no chances and will react, for example, to food scares. Members of the cautious group will consider relevant information but may take a risk depending on the food in question. The noncautious believe that life is too short to worry and will take little notice of warnings (National Consumer Council, 1991).

Customers take account of a range of factors when assessing standards of food hygiene. These include the exterior appearance and management of the exterior and interior, staff working on the premises, the table and its environment, the food and drink, and recommendations from friends. Other concerns are staff personal habits, cleanness and condition of cutlery, crockery, and glassware on the table, and cleanness of any equipment they can see, such as coffee machines. Food contamination, temperature, and freshness are also among customer concerns (Leach, 1996). The principle of the concern is based on cross contamination. That is, contamination passing from one person to another via inadequate cleaning of utensils and toilets, inadequate food handling procedures, and from poor standards of equipment and premises.

As with other images, perception of cleanness arises from physics-based information in the scene. There are three types of dirt. The first is reversible or temporary dirt. This includes dirty ashtrays, unwashed windows and walls, and dust. Also included are fingerprinted, smeared, smudged, and altered menu cards, smeared glassware and chalkboards and stained beer mats (coasters). There are different types of visual appearance problems due to improper wet or polish cleaning. Films, spots, and streaks are deposits caused by water hardness salts or food dried on from a continuous or discontinuous film of water. Deposits that have been wiped cause smearing. Other problems concern the presence of discrete particles of food and feathery white deposits of dried on foam. The detail of the decorative finish significantly contributes to the perception of cleanness. Cobwebs and dust on the bottle of wine was once a badge of quality until the spider farming industry was developed for the purpose of stocking cellars. Once having the appearance of great age, cobwebbed wine bottles are now regarded as merely dirty.

Much of the perception of cleanness arises from the presence of shiny surfaces. Unfortunately, high gloss finishes made from plastic, polished wood, glass, and metal show the presence of the slightest amount of grime. Also, high-transparency materials, particularly glass, reveal dirt. Cleaning occupies a significant part of the time and effort of staff but using good quality fabric that is easily cleaned, strong enough to withstand daily wear and tear, as well as outlawing plain fabric and carpets will ease the task of reducing that unwanted perception of dirtiness.

The second type is permanent dirt. This is dirt that has become ingrained in furniture and carpets, and also includes some physically damaged surfaces, such as dirty, smeared, and chipped paintwork. The mere spot of dirt on a plain table cover, menu, or plain carpet is sufficient to create a perception of dirtiness. Continuous monitoring is required to maintain a customer perception of cleanness. The daily routine includes cleaning, clearing rubbish, and checking the state of repair, efficiency of lighting and ventilation equipment, and removal of animals from the service preparation area. The daily cleaning duties include removing traces of food material and dust on tables and equipment, food debris and smears on furniture and floors, cobwebs on light fittings, and the replacement of dirty linen and aged floral displays. Table cleaning duties include removal of dirty and damaged cutlery, utensils, crockery, and dirty menus.

The third type is dirt included in the decor by the designer. These include use of colors that can be described as *dirty*, and those normally *clean* colors that look *dirty* under some forms of lighting and shadow, that is, in cases where color constancy is not maintained and the visual image of the color changes.

Quality

The perception of quality within the eating environment deserves a separate mention. There are wonderful opportunities for the designer of a quality environment to excel, but there appear to be certain physical rules appertaining to such a space. When layout has been decided, foremost of these rules involve extremes of physical properties such as opacity, transparency, and gloss. Regardless of style, color, and design, quality table linen has a uniform extreme opacity. As cost levels increase, opaque materials tend to be more opaque and opaque linen table napkins carry with them perceptions of higher quality than translucent paper. High-quality glassware in good condition tends to have uniform extreme transparency, extreme gloss with 'sparkle,' and an absence of ridge in the rim. As cost levels increase, materials meant to be transparent must be clean and transparent, increasing the image of cleanness. Quality tableware also possesses extreme properties. Porcelain extremes involve extreme whiteness or extreme color quality. This also involves extreme opacity or the translucency typical of high-quality china. Normally, the higher the cutlery shininess and polish the higher is the quality. Nonuniformity in all these examples implies the presence of dirt. Less than the extreme implies cheapness, age, or, again, dirt. Inevitably, the presence of disposable items made of plastic or paper implies lower quality even though they are more hygienic.

Translucent materials can be used providing they do not look as if they are meant to be transparent or opaque. Similarly, surfaces that look as if they are meant to be glossy should have a high gloss. Materials of intermediate gloss can be used provided that the gloss level appears natural to the particular material, or natural to the material that it is meant to imitate, for example, plastic imitation wood. Criteria of extreme uniform opacity, transparency, and gloss may be different for antique utensils but these are distinguished by shapes and designs characteristic of the period.

It is not necessary that all items in the restaurant be costly. Menus can be expensive, leather-bound handbooks, but the well-executed, artistic chalkboard can be a work of art in its own right, as well as uncostly to create.

There are two types of fake environment. A room may be furnished appropriately or inappropriately with respect to the general layout, age, and aspect of the room itself. A quality environment contains works of art, for example, that are obviously not fake. These need not be expensive, but they must be genuine, they must have the look of not being mass-produced.

Certain venues thrive on fake artwork; modern theme pubs are examples. Everyone knows that the room is furnished and decorated in a kitsch inappropriate manner, but it is enjoyed for its own sake. Customers patronize the new venue that might become fashionable or more likely become a one-off attraction. Similarly, a new landlord in the local pub usually leads to an increase in custom. This accords with findings from industrial psychology. New working practices will initially be more popular and result in higher output. However, new customers will desert if the new atmosphere is not favorable—it is easier to provide what customers do not want.

In the 1960s, 1970s, and 1980s, man-made materials looked like cheap plastic. This is not so true now, but the injudicious use of plastic in a quality environment may reduce the quality effect intended. Plastic in the quality fast food environment, however, carries no such stigma. Light, bright, shiny, plastics have modern, speedy, and, hopefully, clean images unless they have been damaged through cleaning with abrasive materials.

Regarding quality, it is pretence that customers dislike, a venue that pretends it is something it is not. An example is the pretence of charging high prices in an obviously cheap environment.

Some high-cost restaurants seek to extend their intended image of quality by overcharging for smaller items of service and food. Examples are high charges for a glass of mineral water, vegetables that are charged extra to the cost of the entrée, the inclusion on the bill of cover charges, the addition of 15% service charge, and the blank credit card slip entry for an extra tip. These annoyances add to those of over high charges for wine and, in fact, greatly detract from any satisfaction that might have been gained from a high physical and food quality environment.

Comfort

For some customers comfort is the key factor in restaurant selection. Contributing to such feelings are both physical and mental driving forces. Factors affecting physical comfort include squashiness of seating, sound level, and harassment by staff. We are not all squashy chair people; some of us have back problems and need more upright solid chairs. However, a welcome introduction to the eating space is provided by comfortable chairs of correct size for the table. There are furniture designs that have stood the test of time. These are preferable to those compromised by style and simplicity.

Visual comfort also depends on the colors used in the decor. Comfort decreases as:

The number of categorical colors (red, green, brown, etc.) increases, and
The average chroma (saturation or depth) of the colors present increases.

Experiments using computer screens also indicate that the higher the proportion of the red/green component of the color the greater the discomfort. In particular, the greater the red component the greater the discomfort (Sagawa, 2001). It is possible that similar effects contribute to comfort in three-dimensional spaces also. Appetite colors and color awareness are discussed at the end of this chapter.

Mental comfort involves the inner persona of the individual customer. Some feel comfortable only when they are with people, some never relax in company, and some suffer from mental stress when they enter the room. Others are nervous when they find themselves among those who are not members of their own perceived social or professional grouping and/or at a venue in which they feel uncomfortable—perhaps at an establishment outside their comfort zone in the hierarchy sketched in Table 8.1. In the interests of profitability the works of the designer and management together need to be directed towards making their customers feel comfortable and relaxed as quickly as possible. The immediate availability of a cosy seat with the offer of a drink is usually appreciated.

Comfort is a relative perception. Fast food outlets can hardly be called comfortable. However, if shoppers have been on their feet for hours, even a hard plastic seat feels comfortable—but not for too long. This suits the company philosophy, that of giving the customer the food needed as quickly as possible while not encouraging the stay to be over long. For more expensive restaurants, however, the management philosophy may well be to keep customers as long as possible. For that, real comfort must be provided, except when the management seeks to double throughput by turning tables—having more than one sitting per session. The provision of only 'moderate' comfort encourages clients to leave earlier than they might have done. Of course customers always have the right to walk away, unless they are so ill at ease that they must be seen to be eating in such a stressful environment.

Privacy

In a public area, such as a dining room, privacy has much to do with the invasion by others into your space. This invasion takes different forms, but they are centered on the five senses. Some Victorian public houses were designed in a very private way for customers not wishing to be seen. The bar was segmented so that the only person who could be seen, and then only if needed, was a member of the serving staff. No strangers are wanted in some East End pubs with their blacked out windows. At the other extreme are the Witherspoon pubs with their large clear windows discouraging privacy, in the days when it was unseemly for women to be seen dining out their eating area was screened from male diners. Pressure on space now requires that total visual privacy can only be achieved in a separate room.

Privacy for some may be more about not being overheard than not being seen. Contributing to this is the design of spacing and room acoustics, controlled by sound-absorbing soft furnishings within the room decoration. Within a well-designed dining room it is not necessary to invade the senses with piped music. This, particularly if it is loud, robs the brain of the privacy to think. In the packed, youth-orientated public house or disco, loud music provides the ultimate aural privacy, that is, the inability of being able to listen to oneself speak. Rapidly changing patterns of intense light and sound cause saturation of the visual and aural senses, permitting intrusion of sensations from no other sense. The onlooker is swallowed up, dragged in, and controlled.

Privacy from smelling the environment depends on the type of smell. Those arising from damp, drains, stale food, drink, and body odor are not welcome. The way in which we regard being touched differs according to custom. In the crowded pub it sometimes cannot be helped and is generally acknowledged without offence being taken. In the dining room the touch of a stranger is not regarded with favor. On the other hand, in the Italian restaurant, particularly in the south and at festival time, it is impossible to escape the touch of neighboring diners. Physical contact with strangers is discussed in Chapter 4.

In the restaurant, privacy normally depends on a balance between distance from strangers and noise level. In higher quality venues distances are greater and noise levels lower. At lower price venues, distances become smaller but noise levels generally greater, thus ensuring privacy of conversation. When fewer customers are present and sound levels are low the management often increases the volume of music, thereby ensuring privacy is retained.

CHARACTERISTICS AND IMAGES OF EATING AND DRINKING VENUES

Venues and Scales

Over a period of ten months a total of 68 establishments in seven geographical areas were visited and assessed from the point of view of one particular customer (the author). Prior discussions with groups of customers and training exercises resulted in the scales eventually used in the assessments and led to a scoring process that provided as consistent an objective view as possible. The venues, comprising cafes, public houses, luxury and nonluxury bars, and luxury and nonluxury restaurants, were divided by geography and classification as shown in Table 8.2.

This was not a statistically balanced study. Venues were visited as and when opportunity arose and when time was available for a complete scoring of the chosen attributes. All venues were judged for two groups of customer-viewed images discussed in more detail below. The first group includes the viewed properties and general images of the physical environment, including properties of the tabletop. The second includes images concerning the needs of and appeal to the individual and include expectations of the outcome of contact with the space. Each image was scaled from 1 to 9, scaling being carried out from where

TABLE 8.2. Eating and Drinking Places Analyzed by Geographical Distribution

	No. of Establishments					
Location	Cafes	Public Houses	Bars, Luxury	Bars, Other	Restaurants, Luxury	Restaurants, Other
Europe[a]	12	16	0	0	1	0
Argentina	3	0	0	0	3	0
Italy	2	1	0	0	6	4
Miami	2	0	0	2	1	3
Cruise ship	0	0	3	2	1	1
Caribbean	2	1	0	2	0	0
Total	21	18	3	6	12	8

[a]Not Italy

the assessor sat. Awareness colors and prominent features of each space were also noted. Although these results apply to one observer only, care was taken to ensure the scoring was as consistent as possible. The power of the technique in objectively describing an environment is clearly demonstrated.

The Physical Environment

The physical environment was assessed using 24 scales, noted here in italics. Size included a note of the number of dining customers for which seating was arranged, and estimates of space *size* and *ceiling height*, both from low to high. The immediate situation included the degree of contact with the outside world (*enclosed* to *in open air*). Venues having mostly closed windows were usually scored between 2 and 5, those partially or fully in the open from 6 to 9. Also included was a visual estimate of the expected noise level (*silent* to *noisy*) judged from the presence of, for example, loud speakers, musical instruments, and games machines. Illumination was assessed approximately half-way through the visit in terms of the *dark* to *light* and illumination *evenness*, low to high—this included the presence of shadowed areas. Color was scaled in terms of low to high *colorfulness*, *hardness*, and *warmth*. Decoration was assessed in terms of *decorated* to *plain*, the degree of *texture variation*, and the usage of *plants* (low to high), including flowers and foliage in the scene. An assessment of the upkeep of the decoration was also made, in terms of *aged* to *brand new* and *clean* to *dirty*.

The quality of each item on the laid table contributes significantly to the image of quality of the eating place. Tables in the restaurant are ready laid; those in the bar of the public house are laid, usually minimally, just before the food is served. Mills (1989) considered five tabletop attributes. The *tableware* comprises tools and utensils—dinnerware, glassware, and flatware. *Packaging* comprises dinnerware and glassware. The *point of purchase display* is comprised of each individual place setting and table setting. The tabletop as a whole is the *point of purchase display*, which also acts to *enhance the environment* as well as to *market the promise of good food*. Included in the new survey were two other attributes, *napkin quality* and a general quality of the tabletop, *rudimentary* to *sophisticated*. The *kitchen presentation* of the food was also scored. This was judged as soon as the food was received. The meal *cost*, low to high, was also noted.

Individual items contribute to the scoring of the tabletop attributes. For example, there are many types and designs of cutlery. These are well graded, the cheapest being the minimal plasticware, followed by good quality plastic, through tumble-finished cheap 13% chrome in steel cutlery with solid handles. A higher chromium content of 17% has better corrosion resistance. Polished, nickel steel cutlery has a longer lasting shine. The higher the nickel content the brighter and more expensive the cutlery. Hollow-handled cutlery is better balanced than the cheaper solid variety, while luxury market utensils may be made or plated with precious metals. Similarly, condiment sets can range in quality from the cheapest cardboard tubes or plastic sachets, through bright and cheerful ceramics, to solid silver artistic masterpieces.

Cheapest napkins are disposable, of single ply paper and totally inadequate except when supplied in quantity. Some have two, three, or, more acceptable still, four layers. The standard full-size napkins are 40 cm square, but can be found in 25, 33 and 48 cm squares. In higher price restaurants linen napkins should be standard. The good quality, truly opaque, linen napkin well folded, for example, into the rose (ideal for holding bread), or the triple wave (for menu and place cards) can wonderfully improve the look of the table. *Napkin quality* of the paper product was usually scored between 1 and 5, cloth napkins from 6 to 9.

Design and quality of crockery can add greatly to the tabletop impact. Plain or embossed, bold or crested, uniform design or mixed and matched, gray porcelain or white bone china, colors matched or contrasted with the food— each can add or detract from chefs efforts to impress. Quality (and cost) of white crockery depends on weight, translucency, and whiteness. As long as practicality and functionality are satisfied, design can take over and a successful meal can get off to a good start. Poor quality crockery evident to the touch and eye may detract from meal quality.

Ageing of glass can also be judged by surface finish, but quality is judged by thinness, transparency, and manufacturing method. Evenness of transparency and absence of streaks and spots make cleanness obvious. Machine cleaning often results in an increase in apparent dirtiness through the formation of pits that reduce transparency. Washing glass surfaces in hard water leads to disfiguration by deposited mineral and salt deposits. Using demineralized water can reduce these, but only careful hand washing and drying will preserve the total transparency of the best quality glass.

Differently shaped glasses are appropriate for different wines. Smaller, stemmed glasses enable white wine to be kept cool as smaller portions can be served more often. Tall flute glasses for champagne permit the retention of bubbles. A wider, deeper glass for red wines allows the bouquet to develop and be savored. Toughened glassware is one answer to the caterers' breakage problems, but these are not pretty and can spontaneously explode.

Table menus and their layout give a detailed insight into the quality of the establishment. Cleanness and ageing are immediate clues to management attitudes to the dining room. Menu layout and design can increase (or reduce) the cost expectancy of the intended meal sevenfold. At lower-cost eating-places a brand mark shown beside a particular menu item indicates that it has been bought in for resale. This extends the range of dishes available leaving kitchen staff free to prepare the most popular more profitable items. Brand owners hope to create a following, but their products need to be of higher quality than the nonbranded bought-in dishes sold in many of the cheaper venues.

In better-class restaurants the customer will not have a feeling of value for money if the table does not possess extremes of transparency, opacity, and gloss. The sight of highly transparent glassware, which shows wine color at its best, highly opaque table linen, and high-gloss silverware, raise expectations of meal quality. Having cheaper dinnerware, glassware, and flatware is part of the price paid for the cheaper meal. A combination of cheaper tableware and high prices is symptomatic of the overcharging restaurateur—unless the food is of exceptional quality.

Images of the Environment

As a result of previous work on eating environments, the following 20 scales were used in the final survey: *Clear* to *cluttered, private* to *public, rudimentary* to *sophisticated, local* to *posh, tough* to *tender, homely* to *unhomely, comfortable* to *uncomfortable, drab* to *pristine,* low to high *smartness, masculine* to *feminine, uninteresting* to *interesting, pleasant* to *unpleasant, not annoying* to *annoying,* low to high *impact, how the space appeals to me, intimacy* and *elegance.* Expectations of the outcome of contact with the space are summarized using scales of *visual safety, visual satisfaction,* and *visual utility.* Each was scored from low to high.

The scales used are self explanatory, except perhaps that concerning *local* to *posh.* These words are chiefly British expressions and refer to the type of clientele to whom the space seems appropriate. The word *posh* indicates someone who is smart, elegant, or fashionable, someone who is or appears to be comfortably off, a person who might look out of place in a cheaper pub, a place frequented by a *local.*

Venue Characteristics

Although this study was not statistically balanced, there is sufficient data to provide an overall international view of different types of refreshment establishment. Each of 68 venues was scored using the 44 scales. Statistical analysis indicated that the scales could be divided among groups of related images as shown in Table 8.3.

For example, the scales *clear–cluttered,* low–high *intimacy,* and *private–public* were related with an average correlation coefficient of 0.31 to a significance of less than 1%. This group has been named *intimacy.* (The higher the correlation coefficient, between 0 and 1, the better the scales are related and the lower the significance, between 0 and 100%, the more certain the relationship

TABLE 8.3. Image Groups

	Mean Corr. Coefficient	Significance (%)	Group Name
General Images of the Environment			
Clear–cluttered			
Intimacy	0.31	1	Intimacy
Private–public			
Interior: rudimentary–sophisticated			
Local–posh			
Elegance			
Masculine–feminine	0.45	0.10	Elegance
Tough–tender			
Drab–pristine			
Images to do with the Person			
Impact			
Smartness			
How space appeals to me	0.60	0.10	Impact
Uninteresting–interesting			
Not annoying–annoying			
Homely–Unhomely			
Comfortable–uncomfortable	0.49	0.10	Comfort
Pleasant–unpleasant			
Expectations of Outcome			
Visual safety			
Visual utility	0.55	0.10	Expectations
Visual satisfaction			

really exists.) Similarly, other environment images were well related under the group name *elegance*. The environment could therefore be described in terms of two broad images of *intimacy* and *elegance*.

It was similarly evident that images more to do with the person observing the space could be grouped according to *impact* and *comfort*. The third group comprised *expectations*. Some of the following discussion refers to the reduced data, some to the results obtained using the individual original scales. Relationships existing within and among scales and groups are discussed later.

Large differences were found between establishment types. The major difference, predictably, lies in the quality of the decor, the tabletop in particular. For example, scores for *tabletop enhancing the quality of the environment* for the different spaces are listed in Table 8.4. Scores for two types of Boeing 747 British

TABLE 8.4. Scores for *Tabletop Enhancing the Quality of the Environment* for Different Spaces, Means, and Standard Deviations

	Bars, Quality	Restaurants, Quality	Aircraft, First Class	Restaurants, Other	Public Houses	Cafes	Aircraft, Tourist	Bars, Other
Mean	7.7	7.0	7.0	3.8	3.5	2.9	2.0	1.8
Std. dev.	0.6	0.9	—	1.0	2.0	1.6	—	1.0
Number	3	11	1	8	17	21	4	6

Airways aircraft space are included for contrast; these results were not included in the main analyses.

The effect of increased capital and labor outlays on the high-quality environment is clearly demonstrated. This observation was reflected in other attributes dependent on environment quality, such as *smartness, privacy, cleanness*, and *pleasantness*.

The Physical Environment

A correlation analysis of the physical environment scales indicates that there are seven interlinking groups of data. These are concerned with size, lighting, decoration, cleanness, tabletop properties, kitchen presentation, and actual cost. Each is considered separately, the correlation coefficient being quoted in brackets. All coefficients quoted are highly significant, at least to the 5% level.

Size: Unsurprisingly, large venues tend to be set for a greater *number of patrons* (0.51); they also have high *ceilings* (0.32). The latter is linked with *lighting* properties.

Lighting: Venues having *low ceilings* tend to be darker (0.32), have *uneven illumination* (0.43), and are *enclosed* in nature (0.33); these link with *decoration* properties. That is, segregation of areas using uneven lighting is easier in a totally enclosed room having a low ceiling.

Decoration: *Unevenly illuminated* and *enclosed* spaces tend to appear more highly *decorated* (0.39 and 0.35, respectively), with greater *texture variation* (0.51 and 0.24, respectively). Venues having high *texture variation* tend to have a high *plant content* (0.29) and high *colorfulness* (0.42). An *uneven illumination* (0.30), high *plant content* (0.31), and high *texture variation* (0.50) are linked with high *tabletop properties*. The candles, flowers, and assortment of good-quality materials delivering high opacity, high transparency, and high gloss are features of the well-presented table.

Cleanness: *Clean* spaces tend to be *new* (0.70) and of low *colorfulness* (0.29). The uniformly colored surface shows dirt more easily.

Tabletop properties are significantly linked (0.52 to 0.94). Spaces having *high quality tabletop properties* tend to be *clean* (mean 0.35); they are linked with *kitchen presentation* (0.38 to 0.52) and *actual cost* (0.31 to 0.42).

Higher quality *kitchen presentation* is linked with *clean* spaces (0.31) possessing a higher *texture variation* (0.44). That is, greater care in presentation of the tabletop and the food tend to go together.

Actual cost tends to increase with *kitchen presentation* (0.27), as well as in spaces that are *clean* (0.32) and those that are *dark* (0.20).

Hence, lighting, decoration, cleanness, and tabletop properties have separate but interacting parts to play in the creation of the physical eating environment. The analysis confirms to the designer the more vital parts of the decor. However, the low correlation coefficients present around the food presentation and especially around the actual cost confirm subjective judgements that the diner need not get what the diner actually pays for. The venue management may well recoup capital expenditure laid out on the decor by skimping on food quality and presentation. It is much easier to present good quality food in a better way. The independence of kitchen and front of house leads to the conclusion that good-looking meals are not restricted to high-quality environments.

However, the purpose of the decor has nothing to do with the food itself, but it does provide an ambience in which food can be enjoyed in the company of companions selected for the occasion. It helps to generate in the diner the impression that a possibly large bill justifies the quality of the meal. It also helps to instil the feeling that it would not do to create a scene about overcharging or bad service in front of the type of clientele using the venue.

Venue Images

A study of scores averaged across each classification of venue and their significant differences revealed the broad characteristics of each type of venue in terms of their physical environment, and the images and expectations they induce:

Cafes: The general physical environments tend to be *noisy, evenly illumi-nated*, and possess a low amount of *variation in decoration texture*. The immediate physical environment, the tabletop, consists of a *rudimentary* approach to table appearance via a lower quality *food packaging*, and a minimal attempt to increase *expectations*. These properties lead to the generally negative images that they are for *locals*, are generally less *appealing*, and less *comfortable*. In consequence, *expectations* tend to be low.

Public Houses: The smaller *size*, lower *ceiling*, *enclosed* nature, and *uneven illumination* reflect the physical ambience of the traditional English country pub traditionally derived from domestic architecture. These properties are coupled with *ageing decoration* and a lower use of *plants*. The general appearance of the *tabletop* is rudimentary, although *tools and utensils* may be of reasonable quality. Regardless of the rudimentary presentation of the table, *kitchen presentation* of the food tends to be good. This reflects the growing importance of food quality as a profit earner for the British pub. These properties tend to lead to a *local* and *tough* image.

Luxury Bars: Only three of these establishments were patronized, all on board the cruise liner. All possessed a *clean, new, colorful, dark* environment. Features of the tabletop were of high *quality enhancing of the environment*. This physical environment is one that leads to high *intimacy*, and the *sophisticated, tender* images demanded by a *posh* clientele. Greater *appeal, comfort* and minimal *annoyance* followed images of high *elegance* and *appeal*.

Non-Luxury Bars: All other bars were situated in Miami, on board ship, or on the Caribbean islands. The two island examples were simple wooden constructions built on the beach. Hence, they were to different degrees in the *open air*, and of a *light* nature. This and the build contributed to general images of low *intimacy*, low *elegance*, *tougher*, and of a *local* nature. They were also less *comfortable*, more *unhomely*, and *uninteresting*—except as a source of refreshing fluids.

Luxury Restaurants: Restaurants of this type tended to be *large, silent, warmly* colored, and of high *colorfulness*, of high *texture variation*, and high *plant usage*, and *clean*. They also possessed more sophisticated and all round high-quality *tabletop* and food *presentation* qualities. Environment images were of a *clear, private*, and *tender* space designed for a *posh* clientele. These restaurants were *interesting* places of high *impact*. They generated expectations of *safety*, high *satisfaction*, and *utility*.

Non-Luxury Restaurants: The majority of restaurants situated in hotter countries were in the *open air*. They were also *silent*, their colors tended to be *soft* and *cold* with low *colorfulness*. Decoration was *plain*, of low *texture variation*, tended to be *aged* and *dirty*, but with a high usage of *plants*. Also, they tended to be more *cluttered, rudimentary, unhomely*, and of an *uncomfortable* appearance of low *elegance*, catering for *local* clientele. General expectations were of spaces that were less *safe*, less *useful*, and less *satisfying*.

EFFECTS OF THE PHYSICAL ENVIRONMENT ON IMAGES INDUCED

Wide variations, of course, exist within as well as between space types. These reflect different aims, aspirations, and energies of the owners and managers. They also depend on the targeted population. Changes in expectation induced by changes in specific areas of design can be determined. During the survey it became evident that some, if not all, attributes scored were complex in their make-up. Examples are *privacy* and *cleanness*, both of which are discussed in the section on design.

Particular differences to the environment lead to specific changes in the image induced. These are summarized in Table 8.5. The directions of change of each attribute on selected images are noted together with the significance of the effect. (The lower the percentage significance the more likely the validity of the relationship.)

Physical Situation

Table 8.5 indicates that an increase in *size* leads to decreases in perception of *comfort* and *expectation*, while an increase in *ceiling height* leads to a decrease in *intimacy*. The open-air venue suffers from a decrease in *intimacy* and *elegance*. A *noisy* environment increases *intimacy*, but decreases the quality of all other images. At none of the venues was loud music being played.

Lighting

The venues visited provided examples of *lightness intensity* and *evenness*, ranging from bright tropical daylight of the small outdoor bars of the Caribbean islands to the intimate darkness of the luxury bar used after dinner on a cruise liner. *Uneven lighting* of the luxury bar contributes to increased *intimacy* and *elegance*. The *even* high-intensity lighting of the island bar contributes to less *privacy, elegance*, and *impact*. Both types of establishment are, of course, aimed at the same physical population, playing different roles at different times of the day as paying customer of the cruise liner. One caters to the adventurous, outdoor, athletic part of us; the other to the suave, sophisticated, elegant part. Alternatively, it could be argued that these two types of establishments are not aimed at the same population. An objective of the cruise liner may be to provide something that will be attractive to everyone for some of the time. A fuller account of the quantitative aspects of lighting can be found in Chapter 3.

TABLE 8.5. Effect of Physical Environment on Images Induced

Physical Environment	Direction	Change in Image and Significance of the Change				
		Intimacy	Elegance	Impact	Comfort	Expectations
Small–large	Large	Decreases 5%			Decreases 5%	Decreases 5%
Ceiling height (low–high)	High					
Enclosed–in open air	In open air	Decreases 5%	Decreases 5%		Decreases 0.1%	Decreases 5%
Silent–noisy	Noisy	Increases 5%	Decreases 1%	Decreases 1%		
Dark–light	Dark	Increases 0.1%	Increases 5%	Increases 5%		
Illumination: uneven–even	Uneven	Increases 5%	Increases 5%	Increases 5%	Increases 5%	
Colorfulness (low–high)	High		Increases 1%	Increases 1%	Increases 5%	Increases 5%
Colors: soft–hard	Soft	Increases 5%	Increases 5%			
Colors: cold–warm	Warm			Increases 1%	Increases 1%	
Plain–decorated	Decorated	Increases 1%	Increases 1%	Increases 5%	Increases 5%	
Texture variation (low–high)	High variation		Increases 0.1%	Increases 0.1%	Increases 0.1%	Increases 1%
Plants: low usage–high usage	High usage			Increases 1%	Increases 1%	Increases 5%
Decoration aged–brand new	Brand new		Increases 1%	Increases 1%	Decreases 0.1%	Decreases 0.1%
Clean–dirty	Dirty		Decreases 0.1%	Decreases 0.1%		
Tabletop rudimentary–sophisticated	Sophisticated	Increases 1%	Increases 1%	Increases 5%	Increases 5%	Increases 5%

Decoration and Color

The initial impact of the venue is created by its decorative effects. *Decoration* may have no effect on *intimacy*, but it increases images of *elegance, impact,* and *comfort*. Increasing *colorfulness* increases *class* and *impact*. Soft colors increase *intimacy* and *elegance*, while soft and warm colors increase perceptions of *comfort*. *Warm* colors also increase feelings of *impact*.

Inclusion of *plants* and *texture variation* in design increases the designer's options of changing images. *Plants* are rarely found in public houses and bars, but they are common in both types of restaurant, where they sometimes form an important part of the internal structure of the space. *Plants* increase *impact, comfort,* and *expectations*, while *texture variation* increases all five classes of image (that is, *intimacy, elegance, impact, comfort,* and *expectations*). If close examination is avoided the customer can pretend that the artificial plants in the decor are, in fact, real. However, they can cause aggravation, particular to the regular client, because the plants do not change with time and their use is no substitute for the living, changing authenticity of nature in the raw.

Upkeep and Cleanness

Brand new decoration increases both *elegance* and *impact*. *Dirtiness* has a profound effect on all images excepting *intimacy*. *Decoration age* accompanies lack of *cleanness* and lower quality *tableware* and *kitchen presentation*. In this survey, *cleanness* scores fell into three groups. On average, the luxury bars were the cleanest, followed by the quality restaurants and public houses, while the least clean were the other restaurants and other bars. The subject of cleanness is discussed elsewhere in this chapter.

The Tabletop

The setting of the restaurant table can be almost as important as the meal, because its design and layout has a critical effect on the customer's appreciation of the quality of the space. A table ought to look prepared with care, even if this amounts to being perfectly clean and clear. The table and its setting are part of the room and its napery is part of the contrast and harmony within the room. Color contrast is also supplied by butter, which should look cool but be spreadable, condiment containers, and flowers. Within the tabletop, clutter creates confusion, but oversimplicity can create boredom and nonusefulness. Unusual forms, bright colors, pleasing contrasts, and combinations attract interest. Physical domination due to the presence of tall flowers or expensive candelabra distracts the diner. Even though tables are used under subdued lighting in many restaurants they are laid out under a blaze of bright lights, so that imperfections in layout can be swiftly appreciated.

The extent to which the tabletop contributes to the customer's appreciation of the meal depends on the type of restaurant. There are four types of restaurant, the *convenience* or *speed*, the *casual* or *theme*, the *ambience of dining*, and the *serious food* venue. These venues were described in terms of the relative importance of the tabletop functions described above (Mills, 1989). Scores obtained are listed in Table 8.6. For example, *tableware* is seen to be less important to *convenience* and *casual sector* dining rooms, but the *packaging* is more important to the *ambience of dining sector*. A carefully selected tabletop will promote and market the theme of a restaurant in the *ambience of dining* segment. In the *serious food* segment it is the food rather than the environment that is important to the diner.

Finally, the functional worth of the tabletop as a whole was calculated by first giving weights to each function of the tabletop according to their importance. *Tools and utensils* are given a weight of 5, *packaging* 15, *point-of-purchase display*, 30, *enhancement of the environment* 20, and *marketing the promise* 30. That is, in the total restaurant design package, the *point-of-purchase display* and *marketing the promise of good food* are regarded to be the most important images. The weights were used to modify scores given for each function (as shown in Table 8.7), and totalled for each market segment.

In this way, the functional worth of the table is calculated to be least important to the *convenience* sector and the most important for the *ambience of dining* sector. That is, the tabletop and its components have a functional worth that varies from market segment to market segment. Such an approach can be used to calculate and demonstrate the worth of many factors to total appearance and space comparison. This can aid the achievement of optimal design, and

TABLE 8.6. Degree of Importance (Score out of 10) of Various Factors to Restaurants in Different Market Segments (Data from Mills, 1989)

Degree of Importance in Delivering the Restaurant's Promise	Restaurant Sector			
	Convenience/Speed	Casual/Theme	Ambience of Dining	Serious Food
Tableware	8	9	10	10
Table top	5	9	10	8
Packaging	9	8	10	7
Practical level of kitchen's presentation	5	6	8	10

TABLE 8.7. Calculation of the Functional Worth of the Tabletop
(Adapted from Mills, 1989)

Weight	Convenience	Casual/ Speed	Ambience/ Theme	Serious of Dining	Food
Tools and utensils	5	25 (5 × 5)	25	25	25
Packaging	15	135 (15 × 9)	120 (15 × 8)	150 (15 × 10)	105 (15 × 7)
Point-of-purchase display	30	300 (30 × 10)	300	300	300
Enhancement of environment	20	100 (20 × 5)	180 (20 × 9)	200 (20 × 10)	160 (20 × 8)
Marketing the promise	30	240 (30 × 8)	270 (30 × 9)	300 (30 × 10)	300 (30 × 10)
Functional worth score		800	895	975	890

ensure that funding is optimally targeted towards those factors most influencing customer sensibilities and preferences (Mills, 1989).

In the present study, the six tabletop attributes on the original score sheet were well correlated (mean cc = 0.64). The group comprising the *tabletop as a whole*, that is, as *point-of-purchase display*, as *marketing the promise of good food*, and as *enhancing the quality of the environment* were particularly well related (mean cc = 0.91). This indicates that for studies concerned with comparing very different classes of establishment it is necessary to score only one of these attributes. This is because these properties are greatly dependent on the establishment type, as indicated above in Table 8.5. Hence, from Table 8.6, a *sophisticated* tabletop has, like *texture variation* and *cleanness*, a profound effect on the intensity and quality of our image of any venue.

The less well-related tabletop attributes involve the quality of those individual items purchased for the table—the *tableware quality, napkin quality*, and the *food packaging quality* (mean cc = 0.56). Hence, these items are not by themselves the only factors contributing to the qualities of the tabletop as a whole.

Kitchen Presentation and Actual Cost

Food brought to the table should look as if it has been prepared personally and with care. We associate certain physical characteristics to accompany certain images. For example, the more upmarket a space appears the more we expect things to cost. High quality *kitchen presentation* increases the intensity of all five groups of image (mean level of significance = 0.5%). However, good food presentation and high prices can be found in establishments exhibiting poor quality images.

SPECIFIC INTERRELATIONSHIPS

Although similar images of spaces can be grouped together, as shown in Table 8.3, interrelationships between attributes in different groups occur. Such relationships can be used to derive definitions for particular spaces.

A *posh* space (*elegance* group) is *intimate* (*intimacy* group) (cc 0.39), *smart* (*impact* group) (0.46) and *comfortable* (comfort group) (0.46).

An *appealing* space (*impact* group) is *pleasant* (0.70), *comfortable* (0.51) and *homely* (*comfort* group) (0.64).

Some attributes, such as *elegance* and *local to posh* display complex dependencies and interactions, but *masculine to feminine* is a relatively independent scale.

Elegance and Poshness Images

Physical factors contributing to high *elegance* have an *enclosed* situation (cc 0.35), high *colorfulness* (0.40), high *decoration* (0.35), high *texture variation* (0.63), *brand new decoration* (0.40), high *cleanness* (0.48), and a *sophisticated tabletop* (0.32) consisting of a high quality *table display* (mean 0.46). In other words, decoration contributes highly while size and illumination do not.

Elegance is related to other images. An *elegant* space has a *sophisticated* interior (0.62); it is *intimate* (0.46), *posh* (0.52), *feminine* (0.26), *tender* (0.35), *pristine* (0.50), has *impact* (0.55), is *homely* (0.41), *smart* (0.72), *comfortable* (0.53), *appealing* (0.53), *interesting* (0.55), *pleasant* (0.46), *not annoying* (0.56), and has high *expectations* (mean 0.42). An *elegant* space is associated with good *kitchen presentation* (0.33) and a high *cost* product (0.28). In other words, many factors contribute to and depend upon the image of *elegance*.

A *posh* clientele space is similar to an *elegant* one, but it appears not to need the same degree of *decoration*.

Masculine to Feminine Image

Only three factors of the physical environment contribute significantly to the *masculine* to *feminine* image. A *feminine* space possesses *soft* colors (0.35), new *decoration* (0.31), and is *clean* (0.38). The image of *feminine* is highly associated with *tenderness* (0.77) and with images of a *sophisticated* interior (0.26), *elegance* (0.26), a *pristine* (0.39), *homely* (0.26), and *smart* (0.26) space, and one that is *visually safe* (0.25).

Expectations

Expectations of visual *safety*, *visual utility*, and *visual satisfaction* are genuine independent judgements. For example, perhaps we need a thirst-quenching drink of beer but the only place of refreshment is a cafe serving tea. *Visual safety* of the cafe might be high or low, *visual utility* will be positive because a drink can be obtained, but *visual satisfaction* will be low. In this study, however, they tended to be naturally related (mean cc = 0.55). As an example, a *visually safe* space is *silent* (0.25), has *new decoration* (0.29), is *clean* (0.47), and contains *warm* colors (0.38) and high *texture variation* (0.29). It will have good *tabletop* qualities (mean 0.40). It is an image related to high *class* (mean 0.32), high *impact* (mean 0.48), and high *comfort* (0.39) images. It is also associated with a *costly* space (0.33). A space was seen to be more *useful* when its *colorfulness* was high, it was *dark*, and if the *illumination* was *uneven*. It is seen to be more *satisfying* when it was *small* and *pristine*.

COLOR FOR FOOD AND DRINK INTERIORS

When we select paint for a room we do not buy one color but many. Each paint color has an elasticity (Billger, 1999) (Chapter 3) that depends on the quality of the ambient light, the colors that are adjacent to it, and eventually on the amount of dirt deposited on it. Carpets may exist in at least three areas, perhaps part bleached through excessive sunlight, part worn and dirty by passing feet, and part in near original condition in areas beneath furniture where the colors have been protected. The quality of the light falling onto the carpet or wall transforms one color into many. There can be distinct shifts in hue. For example, in sunlight all colors increase in chromaticness; yellows become warmer and more yellowish, and greenish-blue and blue colors shift towards green. In skylight, pale yellow can appear greenish with increased whiteness or blackness and blue colors tend to look more bluish in hue and chromaticness. Some colors change hue. In sunlight reddish-yellow colors look pink. Presence of shadows increases the range of each notional color. Knowledge of this color elasticity enables designers to understand the outcome of their choice of colors (Hårleman, 2001).

The general environment of places where food and drink are served should be welcoming, attractive, comfortable and associated well with the occasion and with what is to be consumed. The design of the table should complement the food and aim to support or increase the appetite. As far as the use of color is concerned, the internal design of the eating and drinking establishments can be divided into three areas. These are the food display area, the wider environment of the room decoration, and the contact area between restaurant and customer. What is appropriate for one area may not be so for another and using Danger (1987) as a guide, colors appropriate for these areas are indicated in Table 8.8. Obviously,

TABLE 8.8. Uses of Color in Three Areas of the Eating and Drinking Environment

Colors	Display Area	Wider Environment	Contact Environment
Hard	Care	Good	Care
Soft	Cooler are contrast to warm colored foods	Care	Good
Bright	Good	Fast trade	Care
Deep	Care	Intimate settings	Accent
Red	Care	Good	Good
Red–medium	Good especially with cooked meat	Fixtures	Accent
Red–Dark	Care	Decoration	Accent
Pink–peach	Good–surrounds	Canteens	Good
Pink–light	Good–fish	Good	Good
Pink medium/dark	Accent fast trade	Accent fast trade	Accent fast trade
Orange—all shades	Accent only	Accent only	Accent only
Yellow	Care	Accent	Accent
Brown–lighter shades	Good	Good	Good
Brown–medium	Care	Fixtures, fittings	Fixtures, fittings
Brown–dark	Care	Small areas only	Accent
Green–clear	Good–but not meat or bakery	Good	Good
Green–light	Good–especially salads	Good	Good
Green–medium	Care	Good	Good
Green–dark	Care	Small areas	Accent
Blue-green–light	Care	Good	Tabletop utensils
Blue-green–medium	Good	Good	Good
Blue-green–dark	Care	Decoration	Tabletop
Blue–light	Good, especially sea food	Good, not large areas	Tabletop, utensils
Blue–medium	Good	Care	Good
Blue–dark	Care	Small areas in luxury venues	Accent
White	Good	Fixtures, fittiings	Good
Off-white	Care	Care	Care
Gray and black	Care	Care	Care

care must be taken whenever color is used, and in the table 'care' indicates that 'extra care' is required.

A major image of all aspects of the venue must be cleanness, the immediate environment being light, clear, and clean. Care should be taken never to use colors that look dirty under certain lighting regimes, that is, they should be selected under the range of lighting conditions to be used in the venue. General advice on the use of colors is not to use too much color, not to use those modified colors that may cause unpleasant associations, and not to use any light or color that may detract from the appearance of food. Neither the reflected colors nor the lighting should interfere with the appearance of the food or be less than flattering to the appearance of the customer. Lighting is considered in Chapter 3.

Appetite Colors

Studies of colors appealing to the appetite indicate that peak appeal occurs in red-orange and orange environments. These colors seem to arouse the most agreeable sensations in diners. Appeal falls with yellow-orange colors, rises with yellow, and falls again with yellow-greens. Cool green and blue-green are well regarded but purple is not liked as an appetite color. Certain colors are liked only when appropriate to the food. For example, gray, most olive colors, mustard, and grayed tones in general are not liked (Birren, 1969).

Danger (1987) lists the good food colors as red, red-orange, orange, peach, pink, tan, brown, yellow, green, blue-green, blue, and white. The poor colors are purplish-red, purple, violet, lilac, yellow-green, greenish-yellow, olive, gray, and mustard tones. This appetite palette can be followed for table accessories. These rules for appetite colors were generally followed in the venues included in the above survey although purplish-reds have been noted particularly.

Colors can affect our mood. For example, those working in offices exposed to warm colors (red and yellow) tend to suffer from higher anxiety states than those exposed to cool colors (blue and green) (Kwallek et al., 1997). It is very possible that similar effects might be found in eating and drinking venues although no such studies have been reported.

Color Awareness

Although interior colors may be important to the customer on entering the space they are not necessarily so when the individual is concentrating on food or another person. Neither the full interior colors nor impact colors were analyzed in detail but included in the survey were those colors of which we became aware while participating in conversation. To some individuals presence of a particular color in a space may not permit them to carry on a deep conversation. However, the assumption was made that a continuous conversation was in progress and colors lying in the field of nonfocussed vision were recorded. Of all scenes recorded, 51% contained two awareness colors, 42% contained three and 7% contained four. These colors were recorded in the following approximate terms:

Reds, greens, and blues in terms of the pure color, the pure color plus white, the pure color plus black.

Intermediate hues, such as purple, orange, blue-green.

The palest colors, such as cream, in terms of pale.

'Wood' colors, in terms of light, medium, and dark brown.

Achromatic colors, white, gray, and black.

Reds, greens, and blues were evident in all venue types, yellow in some, purple, blue-green, and orange rarely. Reds (24 in total) were more often seen in pubs (53%), frequently as floor tiles and brick or textiles blending with these materials. Greens (18) were reasonably evenly distributed (12–32%) among all types of venue, green plus black often being derived from leaves of plants. Blues (16) were mainly found in bars (63%). Yellows (10) were encountered in bars (25%) and restaurants (16%), venues that were perhaps decorated more frequently. Yellow was the least frequently found major color, possibly because it becomes soiled easily.

Very pale colors (27) were the most frequent awareness colors tending to be present in pubs (58%), where they were often used on ceilings and upper walls. Whites (12) were fairly evenly distributed (5–25%) among the venues; in Britain white with black is a traditional combination used inside and outside of buildings. Grays were sometimes found in restaurants (16%), cafes (15%), and bars (13%), but not in pubs. Blacks were found in pubs (21%), but not in bars.

Browns, mostly as varnished, stained, or natural wood, were often found in bars (100%), pubs (90%), and cafes (54%). They were used in different ways. In bars they were in the form of highly polished, high-quality hardwood; in pubs as lower quality fixtures, such as the bar, chairs, and tables. In many cafes, where low-cost decor is found, wood is generally used because it is a useful do-it-yourself construction material that can be cheaply fashioned into furniture. Wood is a greatly flexible material capable of inducing spaces with many different images, influencing physical environment decoration and upkeep, and environment images of *elegance*, *impact*, and *comfort*.

In bars, the most frequent awareness colors are blues (63%), light browns (63%), and dark browns (63%)—black was not found. In luxury bars, colors, especially browns, could be highly glossy. In pubs, the most observed awareness colors were browns (90%), very pale pastels (58%), and reds (53%). Gray, white, yellow, and blue were almost never used. The most frequently found combinations were very pale colors with mid-brown (44%).

Looking at the colors in terms of hard/soft/neutral (that is, neither warm nor cold) and warm/cold/neutral, the colors were predominantly:

In cafes, hard (41%) and evenly warm/cold/neutral (27–37%);
In restaurants, soft (45%) and neither warm nor cold (45%);
In pubs, hard (50%) and neutral (58%);
In bars, hard (62%) and cold (48%).

In restaurants and cafes, there is a more even use of colors. This is probably due to the fact that no specific colors are traditionally associated with such establishments. Exposed wooden furniture contributes significantly to the traditionally hard color image of the pub environment. In the restaurant and cafe, tables are covered with cloths, allowing the designer to use a wider range of colors in the more flexible wider environment. Hard colors are even more evident in the bars surveyed. The nonluxury bars contained the common wooden surfaces found in the pub. However, in luxury bars the hard color image of wood, reinforced by the use of blue, is considerably softened by the presence of the affluence-inducing image of high wood gloss and comfortable luxurious furnishings. Bars and luxury restaurants are influenced more by fashion colors than are pubs and cafes.

REFERENCES

Anon., 1983, *Dinner is Served*, AA Drive Publications, London.
Aron, J.-P., 1975, *The Art of Eating in France*, translated by Nina Rootes, Peter Owen, London.
Billger, M., 1999, *Colour in Enclosed Space*, Chalmers University, Gothenburg.
Birren, F. 1969, *Light, Color and Environment*, van Nostrand Reinhold, New York.
Brown, P. B. and Day, I, 1997, *Pleasures of the Table*, York Civic Trust, York.
Byng, J., 1934–38, *Torrington Diaries Containing the Tours Through England and Wales of the Hon. John Byng, 1781–1794*, edited in four volumes by C. B. Andrews, vol. 2, London, Eyre and Spottiswoode, 18, 49.
Danger, E. P., 1987, *The Colour Handbook*, Gower Technical Press, Aldershot.
De Chiara, J., Panero, J., and Zelnick, M., 2001. *Standards for Interior Design and Space Planning*, New York, McGraw-Hill.
Finkelstein, J., 1989, *Dining Out—A Sociology of Modern Manners*, Polity Press, Cambridge.
Fox, K., 1993, *Pubwatching with Desmond Morris*, Alan Sutton Publishing, Stroud.
Hårleman, M., 2001, Colour appearance in different compass orientations, *Nordisk Arkitekturforskning* **14(2):** 41–48.
Hine, T., 1997, *The Total Package*, Little and Brown, Boston.

Hope, A., 1990, *Londoners' Larder*, Mainstream Publishing, Edinburgh.
Howe, G., 1998, Pub grub—past, present and future, in: *Culinary arts and sciences*, Worshipful Company of Cooks Centre for Culinary Research, Bournemouth, 25–34.
Hutchings, J. B., 1995, The continuity of colour, design, art and science—Part 1, The philosophy of the Total Appearance Concept and image measurement; Part 2, Application of the Total Appearance Concept to image creation, *Col. Res. Appl.* **20:** 296–306, 307–312.
Hutchings, J. B., 1999, *Food Color and Appearance*, 2nd ed., Gaithersburg, Aspen.
Kwallek, N., Woodson, H., Lewis, C. M., and Sales, C., 1997, Impact of three interior color schemes on worker mood and performance relative to individual environmental sensitivity, *Col. Res. Appl.* **22:** 121–132.
Leach, J. C., 1996, Raising food hygiene standards, *J. Royal Soc. of Health* **116:** 351–355.
McKenzie, J., 1980, The eating environment, in: *Advances in Catering Technology*, G. Glew, Ed., Applied Science, London.
Mills, I. J. 1989, *Tabletop Presentation*, Van Nostrand Reinhold, New York.
National Consumer Council, 1991, *Consumer Attitudes and Behaviour in Relation to Time Temperature Indicators*, PD18/D1a/91, National Consumer Council, London.
Ronay, E., 1990, *Good Food in Pubs and Bars*, Egon Ronay's Guides, Basingstoke.
Sagawa, K., 2001, Visual comfort evaluated by opponent colors, in: *Proceedings of the 9th Congress of the International Color Association*, Rochester, NY, eds Robert Chung and Allan Rodriguez 2002, SPIE—The International Society for Optical Engineering, Washington, 299–302.
Tannahill, R., 1988, *Food in History*, Penguin, London.
Zeithaml, V. A., Parasuraman, A., Berry, L. B., 1990, *Delivering Quality Service*, The Free Press, New York.

9

Expectations, Color and Appearance of Food

EXPECTATIONS

The look of a meal provokes expectations, stimulates or depresses the appetite, and can engender joy, a spirit of adventure, or melancholy. When we eat to savour and enjoy, rather than merely to survive, those extra pains taken to use color and appearance to increase temptation and appetite prior to and during consumption are worthwhile. We respond to the aesthetic nature of color, pattern, and design. However, within each foodstuff appearance has deeper meanings and associations.

When faced with food our expectations are comprised of five broad factors. From the total appearance of the food and its environment we make, normally subconsciously, visual estimates of our expectations (Fig. 9.1).

Based primarily on appearance there is a natural order to foods we eat that depend on our culture and what we individually have grown used to; that is, there are colors, flavors, and textures that contrast and complement each other. Examples are roast lamb with mint sauce, pineapple chunks and ham, the elements of a "Big Mac®" or bamboo shoots with water chestnuts. Some foods are more versatile than others and so may carry a surprise. An example is a pancake that may have a savory or a sweet filling.

Today we are used to seeing a great variety of colors in the street, home, and food. This has a comparatively recent origin for the bulk of the population. Colors were not commonly available for decoration and not many colored foods were eaten. Nowadays and according to our personal habits our food day incorporates a riot of different colors as we partake of breakfast, lunch, and dinner. Different people place different relative emphasis on the structure of the day's food but as part of a meal foods are served in strict sequence depending on the culture. At breakfast in Britain we eat browns and white of cereals, bread, tea, coffee, and milk, enlivened perhaps with a flash of color from a fruit. For lunch or dinner we may follow the less saturated browns, reds, orange, or green soup with browns of

Visually assessed safety: For example, when I eat/drink this will I be safe in mind and body, is it safe to eat this product; also will I remain in control of myself, will I drink too much, and if a stranger has bought me a drink in a pub does it contain drugs?
Visual identification: for example, what is this on my plate, how is it different from what I am used to? Specific identifications are:
Visually assessed flavor, i.e., what flavors will this food have when I taste it?
Visually assessed texture, i.e., what textures will this food have when I eat it?
Visually assessed usefulness: for example, how useful will this food/drink be, will it answer my present needs, e.g., will it give me energy, will it make me feel full?
Visually assessed pleasantness: for example, our answer to the question, how pleasant will this eating/drinking experience be?
Visually assessed satisfaction: for example, how satisfied will I be when I have concluded this meal?

FIG. 9.1. Expectations of food.

meats or pie with the pale yellows or browns of sauces and the greens and oranges of vegetables. For dessert we may have the blues, reds, yellows, and whites of a pudding, or the peach, yellow, orange, or green of fresh fruit. For a separate snack we may have paler yellows and browns of cooked cheese. Each meal involves a range of different colors, perhaps accompanied by a splodge of red tomato sauce. The appearance of products that we will eat for one meal is not commonly suitable for another. For example, the cereals and milk we eat for breakfast will not normally be eaten at dinner. The week also has its traditions, perhaps starting with formal lunch on Sunday, ending with sandwiches and cake for family tea on Saturday evening. For others, a Saturday may draw the week to a close by eating out or with a riot of exotic colors bought in with a Chinese or Indian meal. All meals, whatever the cuisine, are initially defined, screened, and approved or disapproved by the veracity of their visual appearance.

The social setting for the meal makes a difference to our appetite and intake, the presence of friends tending to increase consumption. The types of food eaten depend also on the occasion. The colors and shapes of birthday tea foods are different from those eaten at a wedding feast or funeral wake or picnic. Form and color initially define food type. When form is changed by blending or mashing it is the color and context that provides the initial clue to identity.

We all have our preferences. The combination of bright colors with sweetness is the preference for many children, who can be seen eating with relish cakes, sweets, ice creams and puddings. It seems more difficult for many children to acquire a taste for brightly colored nonsugar foods, especially green vegetables.

An indication of the relative importance of the senses may be obtained from studies on primates. The mere sight of food (without odor) or pictures of food is not sufficient to influence the saliva flow rate (Steiner et al., 1977) although the sight of abstract pictures results in a lower flow rate (Birnbaum *et al.*, 1974). However, the amount of salivation in humans increases when subjects are deprived of food and then faced with palatable food. In agreement with these observations, neurones, which respond when a monkey sees its favorite foods, are not activated by smell, touch, or eating in the dark, nor are they associated with salivation. This response occurs only when the monkey is hungry, decreases as it approaches satiety, but is revived at the sight of another food. This also occurs with humans, for whom the pleasantness of a particular food normally decreases as eating progresses. The response is clearly an important factor in food selection (Rolls *et al.*, 1982). These studies indicate the independence of operation of the appearance sense. Hence, appearance governs flavor expectation as a result of experience and saliva release. Once the food is in the mouth, the fact that it looked especially good does not dominate over the opinion of goodness of flavor. That is, if the taste of a food is poor, a good appearance, although raising expectations, will not markedly improve it. If the lights are out we must wait for the possibly slower processes of smell, texture, and flavor to provide product identification. In doing so we eat without the visual confirmation that the product is wholesome. Appearance sets the scene.

As part of their tenth anniversary celebrations in 1997 the Colour Society of Australia organized an eating seminar lunch. This consisted of white foods only: soup, main course, and dessert. Many diners felt uncomfortable while eating because the choice between food identification and conversation had to be made. Although there was a wide variety of foods available some diners found the occasion 'boring' and Chef John O'Connor stated it was the most depressing meal he had ever created. During a meal, color often combined with shape is used as the primary identifier of food. For equally colored dishes the eater is driven to using flavor and texture for identification, while, for reduced foods, flavor is the sole arbiter. The greater concentration required for this and the lack of flavor reinforcement by color was found to be annoying and disturbing.

Appearance provides the initial definition of the product. It provides a state of expectation, a mental image of what the in-mouth properties should be (Fig. 9.1). These expectations provide the framework for visual criticism of the product. For example:

This orange is too big for commercial success;
Its texture will be dry and tough (judged from the dry-looking skin);
Its flavor will be very sweet (judged from its orange/red color);
This orange will satisfy me as a healthy snack.

In this chapter there are discussions of the principles governing the look of food materials, the physiological value of pigments, and the use of food color in tradition and food display.

FOOD COLOR: THE PRINCIPLES

Mechanisms of Food Coloration

In nature, color is indicative of the stage of ripening and freshness. In cooking, formulation and processing color is sensitive to such factors as baking, roasting, and toasting. We have adapted to appreciate and enjoy different types of food and we have learned the color codes indicating when each can be eaten with safety and with personal preference.

Some color hues occur frequently in foods provided by nature. Green, red, pink, orange, yellow, and purple are common. Blue-green is rare, and no blues exist at all. There are fundamental, natural rules governing the colors of the food we eat. Raw natural foods may be divided into four groups. These are foods derived from green leaves, colored fruit and vegetables, mammal flesh and fish flesh.

Plant leaves evolved to be green by default. Their color arises from the reflection of unwanted radiant energy by energy-absorbing photosynthetic systems. Leaves contain a complex mixture of many forms of chlorophyll and carotenoid pigments. The latter have evolved to assist photosynthetic processes driven by the former and to protect the leaf during this process. The different shades of green of leaf foods probably result from the energy absorption optimisation tactics of the particular plant species growing in its natural environment. Whatever the origins and functions of pigments it is the energy not required for photosynthesis that is the basis of our leaf food color. Over age vegetation exhibits the high contrast dark color caused by decay of the chlorophylls and enzymic browning. This is a useful process, however, as foods depending on enzymic browning for their quality include black tea, dates, prunes, and raisins. Nonenzymic browning provides many of the colors, aromas, and flavors of baking and cooking.

The driving force for the evolution of coloration in flowering plants and fruits has arisen from co-evolution of predator vision, and the reflectance and transmission characteristics of the flower or fruit. Fruit maturation involves pigment changes due to chlorophyll destruction, revelation and synthesis of carotenoids yellows, oranges, and reds), and synthesis of anthocyanins (reds and purples). These changes make the fruit visible against the green background of the leaves and so it can be found and eaten by an animal. In general, monkeys eat yellow and orange fruits, those predominantly taken by birds are red and purple, while fruits taken by ruminants, squirrels, and rodents are dull colored, green, or brown. Each animal species has evolved its own retinal cone pigments, which enable it to search for food efficiently. As well as providing food for animals, the plant is also helped. Fruit is usually carried away to be eaten, and the seeds are dispersed where they are spat out or excreted. Blue fruit is rare probably because it would be difficult to pick out such fruit against the background of the sky, so possible driving forces for a mature blue are reduced.

The third driving force for the color of our food is coloration by coincidence. Into this fall those optimisations that happen to result in a biochemical that is colored. Meat flesh color may be an example. Early in the evolution of life on this planet nature used the basic porphyrin structure for respiration in plants and animals. An example is red haemoglobin in mammals. Blood pigments appear within the organism and are not a primary factor in outward appearance. Hence, meat may be red, not as a result of any energy absorption or vision/coloration survival mechanism, but because haemoglobin is a highly efficient respirator.

Many manufactured fun foods that tend to be brightly colored, such as sweets and desserts, contain artificial colorants.

Color is caused by two broad phenomena. When light falls onto an object it may be absorbed and/or scattered. Pigments absorb light and the material appears colored when the light not absorbed is reflected or transmitted. The color is modulated by the presence of light scattering elements within the material. As light scatter increases, so opacity increases and transparency is reduced.

Light scattering contributes greatly to food appearance. Milk color is caused by the light scattering mechanism of casein micelles. The effect of scatter is vividly demonstrated when adding milk to tea. The tea retains virtually its original concentration of pigment; the marked change in appearance is solely due to an increase in light scatter. Apparent creaminess and strength of coffee are related to its degree of translucency.

Cooking of starch grains causes swelling as water is absorbed and a less opaque appearance results. Cooking of proteins leads to precipitation and a more opaque appearance results. When peas, for example, are placed in boiling water the color immediately changes. Water replaces air in cells just beneath the outer layers of the structure. The refractive index of water is nearer to that of the peas than is air. As a result light scattering is reduced and the color appears a deeper green. The opposite occurs when the particulate size of powders is reduced. As the number of reflecting boundaries is increased, light scattering also increases and the powder becomes paler in appearance. Finely milled sugar has a particle size small enough to, first, be undetectable as graininess in the mouth and second, to yield the optimally high light scattering that leads to the high whiteness demanded in wedding cake icing.

Patterns are controlled by optimization of biological and biochemical properties of growth, nutrient transfer and physical strength. Presence of color patterns in, for example, older varieties of apple are governed by principles of natural evolution. These include optimisation of reproduction and seed dispersal properties. There are regional preferences for apple color pattern. Asian markets prefer striped apples, North Americans tend towards blush, and Europeans towards green. Selective breeding controls color patterns in modern varieties, perhaps for properties other than appearance.

Physiological Roles of Food Pigments

Natural food pigments are important to our well-being and seven major groups are found in biological materials. Some pigments found in plants play a significant physiological role in herbivores and carnivores further along the food chain, but there is still doubt as to whether others have any such function (Fig. 9.2).

Animals must obtain vitamins A, B_2, K_1, and K_2 from plants as they cannot synthesize them. A large part of current nutritional advice is that we should eat as many different varieties of fruit and vegetables as possible each day. Most of these contain natural pigments evolved to work by biochemical action and designed, among other functions, as antioxidants to dispose of free radicals. Hence, the protective function of this group of pigments within plants appears to extend to the herbivores that consume them, probably performing similar functions. The value of a variety of natural food pigments to human diet is indisputable.

Color Changes during Food Preparation

Some biological materials are easy to destroy. Chlorophyll-containing mild-flavored green vegetables such as peas should be cooked as quickly as possible in the minimum amount of water, but over-acid water will destroy the color. The

---

Betalaines, for example, from beetroot, might be used for pollination and virus protection. They have no known physiological function in herbivores and carnivores.

Carotenoids from leaves and fruit are used in photosynthesis and pollination. They act as antioxidants and are precursors to vitamin A, which cannot be synthesized by herbivores and carnivores.

Chlorophylls found in leaves and porphyrins in animal flesh are used for photosynthesis in leaves. They may not be essential as a source for herbivores and carnivores as these animals are able to synthesize haemoglobin.

Flavins are used to quench molecular oxygen in leaves. They are vital to herbivores and carnivores, which cannot synthesize vitamin B_2.

Flavonoids are used in growth control and pollination in plants. They act as antioxidants in herbivores and carnivores.

Indoles, have no known use in plants as melanins are synthesized by herbivores and carnivores.

Quinones such as cochineal or carmine are respiratory enzymes. They are vital to herbivores and carnivores, which cannot synthesize vitamins K_1 and K_2.

FIG. 9.2. Physiological roles of food pigments.

volume of water can be increased for harsher flavored greens such as Brussels sprouts. The color of overcooked greens degrades into grayish and brownish pheophytins that in Britain is not favored. The anthocyanin red colors of, for example, beetroot and red cabbage are preserved in an acid soft water cooking environment. The change of color to a dirty blue or purple can be prevented when using hard water by including cream of tartar, lemon juice, and vinegar. The white flavone pigments of cauliflower and onion are retained in an acid cooking medium, alkali turning these vegetables yellow. However, the carotenoid yellows and oranges of vegetables and fruits, and the carotenoid reds of, for example, tomatoes and peppers are unaffected by the acidity of the cooking medium and will not dissolve in it.

The color of raw meat flesh is caused mainly by concentration of the myoglobin molecule. Beef has a greater pigment concentration than pork and poultry. Genetic inheritance, animal age, stress, and diet also affect the color of the raw meat. The milk-fed calf has a cream white flesh, tinged with pink. It is unable to produce red myoglobin because it lacks dietary iron that it normally obtains from eating grass arid grain. The flesh becomes pink by the time it has been weaned at a few months and reaches an optimum eating texture by the age of three months. At six months it is a rosy red, before it is twelve months old cherry red, and at maturity is dark red. This darkening progresses with age and exercise. Anaemic veal is coarser and comes from larger calves six months old and reared on an iron-free water and milk solids mixture.

In life, beef muscle is red, but depletion of oxygen after death leads to a flesh that when freshly cut is myoglobin purple. When freshly cut the newly exposed surface blooms and oxygenates to the bright red oxymyoglobin. As this ages it oxidizes and becomes the brown of metmyoglobin, but this can be slowed by refrigeration. On cooking, myoglobin denatures completely between 80 and 85°C but the conversion rate is slower at lower temperatures. On further cooking to medium and well-done proteins denature, light scattering increases, flesh becomes more opaque, and the pigments more insoluble. The pigments formed are the brown, denatured globin hemichromes and the pink hemochromes. Also on cooking, surface dehydration takes place, haem pigments are concentrated, soluble pigments migrate, Maillard browning reactions occur, thermal decomposition caramels are formed, and eventually on roasting charring takes place.

Salting or fermentation can preserve meat. The Mesopotamians of 5000 BP used salt from the saline deserts to preserve meat and fish. This salt contains nitrates and borax as impurities that produce nitrite then the nitric oxide necessary for the formation of red complexed haem pigments. Nitrite is an antioxidant that retards spore germination. The Greeks knew of these processes and the resulting redness was recorded later by the Romans. Parma ham is traditionally prepared by lengthy seasoning without nitrates. The stable red color is caused by the formation of a myoglobin derivative, possibly through bacterial action. Unlike fresh meat, ham does not become gray brown when cooked because its red color is stable. Marinated meat loses its color quickly on cooking because acid lowers the temperature at which the myoglobin becomes brown.

Redness of cured meats is an indicator of quality as, on ageing in air or under lights, hemochrome pigments break down into brown and gray forms. Reducing agents can prolong the fresh color. Modified atmosphere packaging or vacuum packaging is used to prolong the display of cured products. Pasteurized sliced ham is normally discolored as a result of photo-oxidation during the first 24 hours of display in illuminated chill cabinets. The discoloration can be eliminated with the use of interactive packaging. This combines the use of oxygen absorbers, with concomitant development of carbon dioxide, and packaging material with a low oxygen transmission rate.

The color of fish flesh is governed by blood, melanin derivatives produced for skin coloration, or arises directly from diet. Darker flesh is pigmented with myoglobin that turns brown when heated or exposed to air. It has a shorter shelf life than whiter muscle. Pink flesh color arises from the carotenoid astaxanthin contained in the diet of crustacea consumed at sea. Carotenoids enter this food chain originally from photosynthesis. Carotenoid pigments that degrade little during cooking cause the pink color. However, during cooking, light scattering increases, causing the depth of color to decrease. During cooking, carotenoids can change color. For example, the blue-green of the lobster or crab shell becomes red, but this redness cannot be taken as an indication of thoroughness of cooking. Raw flesh tends to be translucent; cooking leads to protein precipitation and an increase in opacity. Fish oil color depends on species, but the pigments in all types can become yellow or brown through oxidation.

Brown colors of processed foods derive from browning reactions. Enzymic browning causes the colors characteristic of rotting vegetation, but when controlled is involved in the production of black tea and beer. Nonenzymic browning reactions are responsible for the delicious smells and the brown colors that form on baking and frying.

Color Addition

Human beings have long been aware of the importance of food colorings, which have been added to foods since at least 3700 BP when Egyptians colored their candy. Sugar imported into Europe from Alexandria in the twelfth century was colored with madder, cochineal, and kermes.

Chaucer, in the Nun's Priest's Tale, related the contents of the "slender meles" eaten by the "poure widewe" in the thirteenth-century.

"No win ne dranke she, neyther white ne red:
Hire bord was served most with white and black.
Milk and broun break, in which she fond no lack,
Seinde (singed) bacon, and somtime an ey or twey."

The black and white refers to coarse black bread and milk, the general diet being completed with cheese, eggs, and occasionally bacon or fowl.

The majority of natural colors available to the better-off fifteenth-century British household are available to today's cook (Mead, 1967).

> Yellows and oranges are obtainable from egg yolks and saffron ("For hen in broth, color it with saffron for God's sake!"). In Europe, annatto and chocolate were in use in the seventeenth century.
> Blacks and reds are derived from blood (when killing a swan "keep the blood to color the chaudron"), and saunders (from sandal wood, when making gingerbread "if you will have it red, color it with saunders enough"). Dried fruits were being used for black in 1700. Tomatoes arrived in Spain from the New World in the sixteenth century but the red varieties did not become widely used in Europe or America until the twentieth century.
> Green is made from mint and parsley (for glazing rissoles "with some green thing, parsley or yolks of eggs together, that they be green").
> Brown is obtained from brown bread (when cooking Nombles "color it with brown bread"), and ginger. Braun Ryalle (Royal Brown) was made with blends of saffron, green leaves, and saunders.

There have been three ways in which additives have been used to change or restore the appearance of foods. First, we feel more comfortable if the food we are eating is the appropriate color. For example, butter color depends on breed of cattle and season of production. Grass in spring and summer is high in yellow-orange carotenoids and butter from cows feeding on this is more orange than from those feeding on winter pastures when the butter becomes paler. In this case dye is often added to maintain color levels. In the USA, a tax of ten cents per pound tax was levied on the cost of yellow oleomargarine to avoid unfair competition with butter. To overcome the tax, margarine was sold in the white form, together with a capsule of yellow dye, which was added later in the home. Mint-flavored ice cream is white before green coloring is added, while white chocolate is reported to taste less like chocolate than the customary brown product. Product color also influences our ability to identify a flavor and to estimate its strength and quality.

Second, colors are added for manufacturing reasons. Appearance tends to be damaged during processing and added colors help to preserve product identity. They are also used to ensure color uniformity of products that naturally vary in color, as well as to intensify the colors of manufactured foods such as sauces and soft drinks. They also make colorless foods, such as gelatine-based jelly, more attractive. During shelf storage they help protect flavor- and light-sensitive vitamins by a sunscreen effect and serve as a visual indication of quality.

Third, and illegally, they have been used to dilute expensive food materials. For example, in the nineteenth and early twentieth centuries, pepper was diluted with papaya seed and cocoa with brick dust, coffee with chicory, and ginger with pea flour. Alum and plaster of Paris were used to whiten flour, and flour and chalk to whiten milk. More recently, cherry concentrates have been adulterated with a grape and beetroot juice mixture, salmon has been replaced by cheaper trout, goats milk has contained cows milk, betalaines from pokeberry juice have been used to give wine a more desirable red color and appleless apple juice has been marketed. In yogurt and milk products, cherry has been replaced by a grape plus beetroot combination, while apricot mixtures have been replaced by peaches and apricot aroma.

Historically, food manufacturers used for their products much of the range of poisonous inorganic dyes and colorants then commonly used in paint and wallpaper. Edward Abbott, writing Australia's first cookery book in 1864, showed concern at this adulteration, advising readers to grind their own flour, brew their own port, and avoid bright-colored peppers, spices, sauces, anchovies, herrings, and green pickles. Also to be avoided were colored confections, especially those that were green, blue, or red, since the presence of poisonous salts could be expected (Symons, 1982). The present positive legislation in Britain permitting use of specific additives was laid down in 1965. For permission to be gained for a proposed additive, not only must data demonstrating its safety in use be provided, but also the need for it must be demonstrated.

The process of freshening-up the appearance of raw vegetables by spraying them with water is a common sight in warmer countries, such as those lining the Mediterranean Sea. Although this does conveniently add weight to the produce, its main purpose is to restore turgor pressure lost through evaporation. The appearance of older fish can also be thus freshened. This practice was common, for example in medieval Venice, but it is now prohibited.

In Europe there has been a marked change in consumer attitudes to additives where an initial view that additives are "bad" changed to "E (for Europe) numbered additives are safe to eat." Hence many customers checked packs to ensure that additives possessed E numbers. Subsequently, however, attitudes changed and E numbers are looked upon with suspicion. Products are being rejected simply because E numbers are listed among the ingredients. A legislative device intended to be a safety signal for consumers has turned into an alarm bell. Large and intensively competitive food manufacturers and supermarket chains are, under this pressure, tending to market products containing the minimum additives necessary to produce food to fit their marketing strategies. They are excluding those colorants that, although permitted, are receiving adverse publicity. Negative advertising is tending to replace positive advertising. This draws the purchaser's attention away from other ingredients to what is not there and consequently to what is inconsequential. Consumer concern about colorants has led to the wider availability of colorings from natural sources, or manufactured to be molecularly identical to natural pigments. Some natural as well as synthetic colorants have been linked with adverse reactions in some people. In the generally safe area of processed foods, concern for food intolerance is probably now greater than that for food safety.

Artificial colorants used in staple vegetable, fruit, and meat meal products must be used with care because we are sensitive to the color versus quality relationship. Food presented in a natural form, but not in the color expected, may be seen as of lower quality. The situation changes dramatically with nonstaples and the existence of artificial colors allow us to have fun with food. It seems that we can accept impulse desserts and sweets of almost any color. Even named-flavor icecreams can be given quite outrageous colors yet still be accepted as tasting of that flavor. An example is strawberry ice cream that can range in color from orange brown, through pink, to bright purple. Orange desserts having an intense, almost fluorescent color are eaten without reservation.

Although not a color existing in natural food, even blue is now accepted. In the late nineteenth and early twentieth centuries blue was a warning color through its widespread use for medicine bottles. An exception was wedding cake decoration, in which "good luck" is represented by the colour blue. The medical use of blue has ceased, resistance for its use as a food color has declined and it is now available for use as a dark contrast color in milk products, sweets, and icecream.

Attempts to omit color additives do not always meet with success. A possibly apocryphal story circulating in Sweden in the early 1980s concerned colorants omitted from orange drink. Because cashiers at self-service restaurants could not tell the difference between the orange drink and plain water, colorants had to be reintroduced. Another unsuccessful colorant omission concerned clear colas.

Some colors are unwanted and spoil product appearance. Salted fish darkens with time but this, as well as bloodstains, can be bleached out using hydrogen peroxide. Marinated herring, tripe, and in-shell walnuts are lightened similarly.

Color Associations

Color and appearance are powerful indicators of object quality and this applies particularly to food. Human beings have different sensitivities to flavor and it is relatively easy to confuse tasters by giving them inappropriately colored foods. Particular colors are associated with particular flavors and up to certain levels the deeper the color the stronger the associated flavor. The effect of a further increase in color depth depends on the food material; tomatoes appear over ripe and past their best, canned meats appear to be artificially colored and of poor taste. Color affects other sensory characteristics, and hence food discrimination, through learned associations. For example, if the cherry redness of a cherry drink is increased, the perceptions of cherry flavor and sweetness increase. That is, when color vision skills evolved to find and identify food, they also evolved to characterize color/taste/flavor associations. This is because redder fruits are normally riper, more edible, and sweeter.

Food appearance influences expectations in different ways. In a trial on vegetable protein frankfurters, two populations of taster were found. One group preferred pale colors because saturated colors seemed not natural. The other group preferred saturated colors because pale colors suggested a high fat content (Noel and Goutedonega, 1985). In the United Kingdom such divisions within the population include preferences for tomato soup color. One distinct group prefers orange-red soup, the other blood or cherry-red. The first group had been brought up on Heinz Cream of Tomato Soup; the other group was more used to soup made from tomato powder or puree.

Our perception of in-mouth assessed flavor can be affected by many factors, including sample appearance, its environment, the attitude of the person asking us to taste, from our own physical environment, as well as the mood we happen to be in at the time (Chapter 1). Preconditioning also affects us. Panellists, who are acutely aware that high fat content foods are bad and also aware that they are tasting a high-fat product, will normally perceive such a product to have a worse flavor than a product seen to be low in fat. There appears to be an age dependency on tasters, older subjects being more sensitive to visual cues. We are also influenced by brand. If we are aware of the brands of beer we are tasting, our regular brand will always taste better. This is because a product brand does not only have specific physical and flavor properties, it also has cultural and emotional images.

Field-independent tasters attend to their taste and smell perceptions when classifying flavor without regard to what may be an inconsistent visual stimulus. *Field-dependent* subjects make more mistakes when trying to identify flavors in the absence of visual cues as to their origin (Moskowitz, 1983).

Advertising is a complicating factor. For example, the British have been subjected to the persistent advertising claim that "smaller peas are sweeter." An unwary approach to the panelling of peas resulted in erroneous findings founded upon knowledge of this claim.

FOOD APPEARANCE: THE PRINCIPLES

A food product or material can be described in terms of its total appearance (Chapter 1). The total appearance of a food consists of:

Visual structure: for a piece of bacon this will consist of definable areas and shapes of the different muscles, fat, and gristle, each element of the visual structure possesses a

Surface texture (such as the fibers in meat), as well as

Color, Gloss and Translucency and Patterning of color, gloss, and translucency.

Temporal properties: such as ageing or the dynamic display of the wobbly jelly or the waiter's preparation of the flambé.

These properties, together with the illumination by which the object is lit, govern the stimulus for us visually to identify and interpret our immediate environment. Using these properties we can define the look of the object. For each object that we recognize we have within our memory critical dimensions for each of these total appearance properties. Separate sections are devoted to visual structure and surface texture, translucency, gloss, and appearance ethics. The dynamic display is considered further in the section on food display.

Visual Structure and Surface Texture

Structural patterns in natural products are characteristic of a food and are among the means by which it is recognized. In composite natural foods we demand the correct proportion of each element of the structure. Significantly more beefsteaks and roasts are purchased when fat thickness is less than 8 mm. For greater acceptability, lamb chops cut for display are trimmed to give specific lean to fat to bone ratios. In composite manufactured foods, such as pizza, a complex surface pattern of ingredients is possible. However, the appearance of each constituent of the matrix must fulfil the criteria of freshness or goodness normal for that constituent. Very precise and regular patterning tends to give the impression that the pizza has been manufactured by machine and the look of perfection betrays the factory production line appearance signature of super-market mince pies. Evidence of hand making in decorated chocolates is welcomed. Exceptions to this hand-crafted appearance occur at a restaurant that has a highly skilled chef. Formed products made and served there are expected to look perfect, not rustic in nature.

Perfection of appearance is the shopper's main criterion for selection of fruit and vegetables in the supermarket. For example, for wholesale and retail markets it is desirable that apples in a box are presented in the same way and look as similar as possible. Our preoccupation with appearance has led to the rigorous standardization of size and shape and a reduction in the number of varieties on sale. However, items of imperfect appearance are restricted to 'organic' produce.

Food Translucency

A translucent material is one that both transmits and reflects light. As a phenomenon, translucency occurs between the extremes of transparency and opaqueness. Transparency or clarity, like window glass, occurs when there is no visually apparent light scattering, it is the transparency or opacity properties that tell us whether a food sculpture is made from ice or lard. Transparency is a property often sought by the processor. Ancient Egyptians used layers of stones and shellfish shells, and wool and cotton fibers as filters to clarify water and drinks. The haze in spirits, wines, and beers normally leads to their rejection, although in Germany and Belgium there is a preference for the whitish-yellow hazy wheat beer. This product is naturally translucent as wheat lacks the husks that create a natural filter. Anisette-based drinks are clear when poured from the bottle, but turn white when water is added. For their flavor these drinks rely on aromatic terpenes, which are soluble in alcohol but not in water. When water is added they are forced out of solution to form a milky haze. Addition of milk to coffee or tea decreases clarity and increases opacity.

Product identity can depend on the extent of the clarity. Clearer orange drinks may be seen as long, refreshing, probably containing artificial colorant, and more likely to be appreciated by younger people. More opaque, lighter-colored fruit drinks, on the other hand. are often regarded as health-giving breakfast beverages obtained from real fruit and more likely to be appreciated by adults. Turbidity in fruit juices can be a positive or a negative attribute depending on the expectation of the consumer. In origin, hazes are biological, caused by growth of microorganisms, or chemical, caused by physicochemical instability and the coagulation of particles that tend to form a sludge. Opalescent apple juice contains a stable cloud of soluble as well as insoluble pectin-stabilized particles. Loss of clarity in fruit conserves such as jams may denote the inclusion of pectin or corn flour as thickeners. Custard and sauces that have been made using water are less opaque because those that are milk-based scatter more light. Gelatine-based jellies have greater visual impact if they are clear. The albumin of fresh duck eggs develops into a turbid gel, but immersion of the eggs in sodium hydroxide and salt results in an increase in transparency.

The color of raw salmon flesh is a translucent deep pink-red, which on smoking turns a more opaque light pink. As the conformation of the protein changes during processing light scattering within the fish increases. This has the effect of reducing the amount of light penetration so giving the pigment less opportunity to absorb light. Hence, as well as becoming more opaque, it is a paler color. The degree of light scattering within a structure governs the perceived depth of color or chroma of flesh foods, for example, as a result of cooking. The appearance of a smoked salmon slice is due to the interaction of natural pink pigment, imposed browning, and the translucent nature of the flesh. Hot-smoked salmon is cooked as it is smoked. This process causes an additional increase in opacity and the delicate translucent effect is lost.

Similar effects occur in fresh meat, particularly pork, that can be unduly pale in color. Under certain pH conditions, sarcoplasmic proteins are released and precipitated, light scattering is increased, the effect of the pigment present in the muscle is reduced, and the meat looks paler. Freezing rate also affects the perceived color of foods. Faster freezing leads to smaller ice crystals, increased light scatter, and paler beef.

The color of green vegetables immediately changes when they are plunged into boiling water. This occurs because water replaces air in cells just beneath the outer translucent layers of the structure, reducing light scattering and intensifying the depth of greenness.

Much of the cooking process is aimed at changing food texture through a change in structure. Protopectin is solubilized, cellulose is softened, starch takes up moisture, and proteins are precipitated. All these factors contribute to the perceived color, mainly through changing the translucency. Light-scattering phenomena also contribute to powder reflectance properties, again color changing with particle size. The smaller the particle size, the greater is the surface area, the higher the reflectance, the paler the color.

Tea brewed in hot water develops an unsightly surface film or scum, which increases scattering. This scum consists of calcium carbonate and an organic matrix insoluble in all solvents except highly concentrated alkali. The scum forms in water containing both calcium (or magnesium) and bicarbonate ions. It is greatly increased by the addition of milk, but no scum forms in lemon tea because of its high acidity.

As part of the display of food in bulk and on the plate the dimension of transparency is provided with ice. Small ice carvings can be used to decorate dishes of pate and mousse. Large sculpted blocks are used for adding a height dimension to a display. Crushed ice is used to show off fruit such as melon and avocado as well as shellfish and wet fish.

Transparency and translucency are brought to the meal through consommé, sauces, aspic, and gelatine. Some authorities see transparency as desirable because it indicates purity of ingredients and careful preparation. However, since the advent of bland-flavored colorless waters, launched in a bid to extend the bottled water market, clarity may be seen as bringing more artificially and unnaturalness to foods. Similarly, some see clear apple juice as unnatural because it is a naturally cloudy product. Similar is the belief that white signifies high purity. This led directly to the overrefining of sugars and the use of bleach to whiten flour for white bread.

Food Gloss

Light scattering also takes place from the surfaces of materials, giving rise to perceptions of gloss. When we look at a surface it is the scattering pattern with respect to the angles of lighting and viewing that are responsible for the perceived degree of gloss. Particular gloss characteristics are associated with different fruits and vegetables. Surface scatter is more diffuse from rough surfaces, such as an orange skin, than from smooth, such as aubergine (eggplant) and unripe tomatoes and bananas. On the plate the gloss of the burned surface of a rare steak gives notice of its rareness. This gloss is different from that of the burned surface of the well-cooked piece of meat.

High-quality chocolate normally has a high gloss and light scattered from the surface is near mirror-like specular reflection. When chocolate blooms it loses gloss, the specular reflection changes to diffuse scatter and the surface becomes dull. This lack of gloss in chocolate is symptomatic of the state of crystallization and granularity. The three types of bloom can be induced by melting and recrystallization, or by storage at below melting point temperatures, and/or by migration of non cocoa butterfats.

A wide range of gloss occurs in foods, but extremes are rare. Gloss is symptomatic of production method in yellow spreads, some of which achieve a relatively high gloss. Gloss difference helps us distinguish between components of mixtures such as fat floating on hot gravy. Glossiness of moist surfaces such as fish reinforce perceptions of freshness, but only if there is sufficient directional light. Under diffuse illumination fresh fish appears 'dead'.

Waxy layers and deposits are components of the surface cuticle tissue of many fruits. As well as preventing fluid loss, their natural gloss reduces penetration and damage from visible and ultraviolet light—shaded plants rarely possess surface gloss. Different varieties of apples exhibit different levels of gloss. Glossy surfaces look attractive. Hence, wax additives designed to reduce gas exchange, weight loss, and fungal growth are also designed to be glossy. These include fruit waxes such as carnuba wax, polyethylene wax, beeswax, rice wax, and shellac.

Gloss or glaze has three meanings in the kitchen: to finish vegetables glacé, to color under a salamander as with fish in a sauce, and to coat with a jam or fruit purée, as with a flan or tart. Dried foods and baked foods tend to be matt if not deliberately made shiny with perhaps a sugar glaze. Freshly cooked vegetables are coated with butter so that their shininess is preserved while the meal is being eaten. Sweet or savoury aspic glazes are used to decorate a wide range of foods; in Spain these include the succulent, suckling pig for the table.

Gloss is added to display by low temperature spotlights shining onto wet or glossy surfaces and ice. Frosting the rim of a glass removes the bulk gloss of glass and substitutes sparkle. Multicolored edible glitter on a gum arabic base can be used to decorate low water activity cakes and icing.

An opalescent or iridescent sheen sometimes appears when meat is cut transversely to its fiber axes. This can occur with raw, cooked, fresh, or cured meats and, when viewed from a particular direction by the naked eye, is green or less commonly red. This is an interference phenomenon believed to arise from reflections from layers within individual meat fibers.

Food Appearance Ethics

Included among ethical problems are first, where a particular customer belief is being exploited to the gain of the manufacturer or marketer and to the cost of the customer, and second, where the customer is being misled over the quality or origins of food being purchased or consumed.

One of the reasons for brand loyalty is that the customer has developed a trust in the ethical principles practised by the supplier. We tend to trust what we think we know and believe that someone out there is looking after us. It is to the producer's advantage to foster this belief not only to increase sales but also to avoid the losses that result when we find out that our trust has been betrayed. This transfer of trust when we visit some countries leads to illness when we are tempted by the succulent appearance of fresh produce or unbranded ice cream.

Added colors can present ethical problems. Obviously the manufacturer must abide by the additive regulations of the country in which his product is sold. However, there are ethical considerations not specifically included in statutory regulations. One concerns the nutritional education example that fruit juices are a good source of vitamin C (ascorbic acid). Unfortunately this chemical degrades anthocyanin pigments, a major source of fruit color, and therefore the manufacturer would prefer not to add ascorbic acid. However, not including it constitutes a deliberate undermining of good nutritional education. It is professionally unethical to market fruit juices containing little or no ascorbic acid. An even worse practice is to fortify fruit juices with riboflavin and thiamine and label them fortified. The public has come to accept the word as meaning fortified with vitamin C.

Yellow additive can be used in cake products to give the impression of a greater egg yolk content. The manufacturer may logically argue that differences in yellowness can be caused by some yolks being a stronger color than others, or that it is an assumption that the consumer is concerned about the inclusion of eggs anyway. Similar to the fruit juice example, is the manufacturer deliberately trying to induce the impression in the consumer's mind a fact about his product that the manufacturer knows is patently not true? After all, egg yolks contain factors that are essential for our well-being. How much is the manufacturer consciously or subconsciously relying on a customer belief that others are looking after their welfare?

Whereas adults are attracted by more subtle shapes and colors of the beautifully boxed chocolates, children are attracted by bright colors. Products targeted at children in the United Kingdom include Alcopops—lemonades containing alcohol. These are marketed in brightly colored bottles with brightly colored labels decorated with near cartoon characters, media heroes, or portraying someone who could be under 18 years old. Under public pressure some have been withdrawn from supermarket shelves. These products have probably led directly to an increase in alcohol abuse by children and young adults, but who is to blame? Perhaps it is the directors of the companies seeking to fill this unethical 'market niche'. Perhaps it is the designer of the bottle decoration who is not behaving professionally? There are similar problems in the USA where manufacturers are deliberately targeting children with highly colored high-sugar and/or high-fat content products. Like the sweet and candy manufacturers they seek to increase consumption of worthless obesity-creating foodstuffs.

The misleading menu presents many opportunities for deliberate or accidental deception, one fortunately pursued by trading standards officers. In some areas of Britain one in five menus are flouting the law in some respect. Products labelled by the manufacturer as *reformed, chopped and shaped,* or *with added water and gelatine* are described on the menu as *scampi, steak,* or *gammon.* Descriptions such as *scampi, steak,* or *gammon* refer in law to the use of an unprocessed product in the dish. The word *burger* is illegally used for products containing less that 80% meat. It is *economy burgers* that need contain only 60% meat. A meal described as *roast* should be roasted, not made with meat that has been steamed and flash-roasted. A *smoked* product should be smoked, not contain added smoke flavoring. Such a product labelled *smoked salmon* ought to be described as *smoke-flavored salmon.* Products made with cheese substitutes should be labelled *cheese flavored. Crab salad* is illegally used to describe salads containing crabsticks, which contain little crab. *Fresh fruit salad* should not contain tinned fruit. *Fresh* should mean food that has not been frozen, processed, or adulterated. Care is not always taken with product origins, *Scottish salmon,* for example. *Home-made* food should be made on the premises, hence *home-made soup* should not contain packet soup. There is a good chance that in the restaurant the quality of bought-in food is lower than that made on the premises. Almost inevitably the closer the ties are between vendor and restaurateur the better the quality to the customer.

Displays can be misleading also. An example is the choice, low-fat, low-gristle slice of bacon on top of the pile that covers fattier less desirable pieces beneath. This is similar to the market stall technique of showing the most perfect fruits and vegetables in the front and serving the customer from the back of the display from a lower quality selection. Another ethical display problem is the illumination of the fresh beef counter with red-biased light. This conceals the brown specks that indicate pigment oxidation; the customer normally regards such meat as undesirable.

FOOD DISPLAY: THE PRINCIPLES

A well-presented display of good quality food in a suitable quality environment provides a substantial stimulus for purchase or consumption. The better we are fed the more important color and appearance are and the more appealing to the eye the food must be. There is no sadder and more depressing sight as a plate of dried-looking sliced meat, stale sandwiches, or a dilapidated sweet trolley carrying plates of largely demolished cold desserts.

Presentation and visual merchandizing are vital elements of the ambience of both dining room and store. The principles have long been known. Wherever it is, the attraction of a display or a presentation for selling involves imparting a visual impression of a clean, clearcut, restrained design, incorporating good lighting, color contrast, originality, tidiness, symmetry, and a judicious use of space. The advertisement, as opposed to the display, will also possess a brief legible message.

When you are hungry there can be no more mouth-watering sight than a well-laid-out buffet or table display or cool cabinet of fresh-looking sandwiches and bottled drinks. These are the factors that drive us to eat.

Tradition of Food Display

The more expensive occasion has always been the opportunity for a chef to display creative artistic talents. In the foureenth century, animals were displayed in their natural state for the feast. Swans and peacocks were skinned unplucked, then roasted and served reassembled with feathers and gilded beak. Elaborate creations called 'subtleties' made from colored marzipan or pastry were also displayed at the banquet table for the amusement of guests. These may have included animals and birds or religious or battle scenes. Among early fifteenth-century practical joke subtleties were cooked meat that was made to appear raw, meat that appeared to be full of worms and pots containing soap that appeared to boil over. Most subtleties were made of sugar, some of pastry, and some of ground meat. A 'wyf lying in childe-bed' was a suitable subtlety for the wedding feast. Rather more down to earth is the modern modestly clothed figures of bride and groom on the wedding cake. For the traditional wedding it may be a towering white cake that provides height to the display. An early nineteenth-century pastry cook, using the steeple of St. Bride's church, Fleet Street, as a model, originated the tiered and pillared wedding cake.

At the medieval banquet the 'great salt cellar' was likely to be the major item of table decoration. It was frequently made of silver and of splendid design, possibly featuring armorial bearings or made in the design of a ship.

In Elizabethan times pasteboard pies containing live frogs or birds were constructed. Brought to the table was a paste stag with an arrow in its side. When the arrow was removed claret flowed from the wound. At the merchants annual feast there were "geliffes of all colors, mixed with a variety in the representation of sundry flowers, herbs, trees, forms of beasts fish fowls, and fruits, and thereunto marchpané (marzipan) wrought with no small curiosity, tarts of divers hues and sundry denominations. . . ." The natural colors from flowers and leaves were used. A favorite dish of the old French royal family was the lièvre à la royale, a dish made from hare. For presentation of the dish the skin of the hare was stuffed and seated on a throne in the dining room with a crown on its head.

The principle of these displays was to include the spirit of the hunt in the dining room and incidentally identify to the eater the origin of the dish. In modern times display of a recognizable part of the animal is rare and use of a boar's head, for example, can be found only as part of a ceremony that celebrates the past. Animal origin tends to be disguised, perhaps in the form of fish fingers.

At the nineteenth-century banquet in France, cooked meat or poultry was displayed on an elaborate plinth. This was made from a mixture of kidney and mutton fat or moulded with butter fat, lard, or wax. Architecture and flavor were closely connected, the plinth being an integral part of the prepared dish. Among the many designs of plinth were vases with double handles, plinths with birds, a crown, an angel's head, dolphins, and vine leaves. Meat was carved in front of the customer, enabling the appetite to be enhanced by the sight and aroma.

Anton Carême was a master of early nineteenth-century gastronomic display. This took the form of elaborately designed dishes in which foods were made into pieces of architecture such as small-scale pagodas, fountains, and temples sculpted from sugar and wax. These sometimes stood four or five feet in height and were displayed in the center of the table. They were the forerunners of the modern centerpiece of candles and flowers. In France, huge bouquets hid guests from each other. These were later replaced by short-stemmed flowers and larger candelabra to increase light levels. Decoration declined further leaving the table comparatively clear. Unfortunately this has resulted in tables being reduced in size, thus increasing levels of clutter and discomfort.

Food color changes with time. In ancient Rome, for example, wealthy hosts would order relays of waiters to run more than 30 miles from the port of Ostia to Rome carrying red mullet in bowls of unchanged water. These would arrive at the table in time for those present to see the gold, purple, and blue scales fade as the fish slowly died.

Physical actions involved with serving introduce a dynamic into the display and increase the diversity of appearance. The way the waiter serves the meal can make or break an occasion and the display of the chef as the flambé is prepared is part of the meal. In first-century Rome, the tenth-century Byzantine court, and thirteenth-century Far East, ingenious mechanical devices used the power of water, steam, compressed air and counterweights to provide means of serving guests. The "four and twenty blackbirds baked in a pie" was a seventeenth century method of creating temporal diversity of appearance at the table. This "revelry pie" was blind baked and consisted of a real meat pie inside an empty pie crust. Live birds or frogs inserted into the cooled pie through holes in the base were released when it was cut open.

The most well-known badge of the quality cook, Le Cordon Bleu, was established subsequent to the founding in 1578 of L'Ordre des Chevaliers du Saint-Espirit. Members of the order, identified by wearing a gold cross on a blue sash, set high standards for themselves and those who served them. These cooks tied their aprons with blue ribbons and hence the link between the color and culinary excellence was established. By the middle of the seventeenth century, France had established a cuisine of its own and the first restaurant was opened during the reign of Louis XVI. In 1792 he, with other members of the aristocracy, were guillotined. This released cooks from the great houses onto the streets of Paris, where restaurants began to flourish.

The multidimensional appearance nature of the table is illustrated by descriptions of the Australian eating scene. In 1870s Melbourne the "sixpenny restaurants shone with plate glass, white linen and pretty waiter girls," while Pfahlert's Grill in 1928 Sydney, "had its benapkined rounds of stilton cheese, crystal bowls of crisp celery, porterhouse steaks, pewters of foaming ale" (Symons, 1982). In other words, there were the basic perceptions of color and color variation, transparency and transparency variation, opacity and opacity variation, as well as visual structure and surface texture variation. The scene also exhibited the derived perceptions and expectations of clear visual identification, and visually recognized cleanness, flavor, texture, and satisfaction.

There has long been a history of cleanness at table. In medieval Britain tables were covered with 'fayre and clene' cloths. In the best households it was made of the finest linen. Lesser members of the household had lower grade or older linen or nothing at all as they ate at a bare wooden board. The cloth was laid by two members of the staff, one using a rod to smooth out the wrinkles while both stretched the cloth between them. Linen napkins were used. It was generally expected that the table be kept as clean as possible during the meal.

Elements of Design

Static as well as dynamic display form an essential part of entertainment on a large or small scale through the physical arrangement of food and decoration and the behavior of the serving staff. The elements of a table display are similar to those of art advertisement and the shop window. They consist of the motif, the organizational plan, and the layout.

A display frequently uses props to set the food within its product, appearance or color themes. For example:

Bread is displayed with baking trays, rush matting, and brass;
Dairy produce is seen with milk churns, fresh straw, and marble;
Wine and spirits are shown with casks, antique furniture, and wooden panels;
Fish is set off with lobster pots, ceramic dishes and figures, and translucent pearlized plastic; and
Icecream and summer drinks can be shown with fake cream, patio tables, sunglasses, and fake fruits.

The units of a pattern or design motif tend to be of three general types. These are motifs based on natural shapes, geometrical motifs based on the pure forms of the circle, rectangle, and triangle, and abstract and decorative motifs, which do not look like natural objects. They may be arranged to depict a character, such as dignity, pleasantness, or restlessness. The character of the motif is determined by the direction of lines and by the shapes, sizes, and colors of the motifs and spaces.

The overall design has an organizational plan consisting of a number of elements (Emery, 1980; Radhakrishnan, 1980). The plan consists of design elements combined to achieve a particular effect. This may be performed in a number of ways (Fig. 9.3). This organizational plan is created into a display using line, color, and mass, within a series of broad geometries, (Fig. 9.4).

The discipline offered by such techniques of formal analysis has obvious advantages for many products and industries. Relevance to the food business is wide, from packaging, to store design, to the restaurant, the window display, to the construction of compound food products, and to the design of buffets and table layouts.

Alternation, in which the basic shape is repeated as it undergoes changes in organization, size, and color.

Harmony, in which units are harmonious when one or more of their qualities such as shape, size, or color are alike. Extremes are complete repetition and discord. Harmony is in between, and contains elements of both.

Contrast concerns the bringing together of two opposite qualities in shape, line, or color. Types of color contrast are discussed in Chapter 2.

Dominance, involves the proportion by which some parts of a design are given more importance.

FIG. 9.3. Design organization elements.

Curves and waves have a correspondence of line or contour. They inject a rhythm or flow into a display and imply movement or stillness, and calmness or discord. Age and decay is depicted when lines are weak.

Radiation occurs when a subordinate line arises from the center or from a major line in the display. It may occur as spokes of a wheel from the axle.

Mass, through shape, form and color, gives rise to perceptions of symmetry, rhythm, repetition, grouping, and diversity.

Repetition, which may be single items or identical groups arranged in straight lines, diagonally, or curved.

Within the overall display there is a balance. Balance is a visually perceived coordination of mass, form, color, and area. Formal balance is a symmetry of area or weight about an axis running at any angle through the design. Informal balance is one in which the perceived balance occurs even though there are different units of mass placed each side of the axis. Factors governing the choice of balance are the focal point, points of interest, and lines of sight.

FIG. 9.4. Geometries of design.

The visual sense is heavily involved in our appreciation of the meal as a whole and the design in particular. The preparation and serving of plated meats is judged professionally according to a number of attributes (Fig. 9.5).

Vision and appearance are used in the scoring of all presentation and all workmanship attributes, most composition, much of the creativity attributes, and all the hygiene factors.

Presentation, including food freshness, correct portion sizes, correct ratio of garnish to portion served, color, appearance, use of proper serving dishes, the overall cleanness, and attractiveness of presentation.

Workmanship, including cleanness of cut of serving, correct aspic, coating of items, the correct preparation of ingredients and cooking, correct order of work and professionalism.

Composition, the color, texture, and flavor combination, the harmonization of accompaniments and functional composition.

Creativity, innovation, including the presence of new ideas and combinations, use of different garnishes and degree of difficulty and balance.

Hygiene, all aspects of which are taken into account.

FIG. 9.5. Judging the preparation and serving of meals.

Creating the Show

If no-one eats, the cook is out of a job. It is the buffet table that acts as a major showcase for the creative chef and inspires the appetites of those partaking of the feast. Shape and color can be varied. Radishes, turnips, potatoes, and beetroot become flowers; pimentos, hard-cooked eggs, colored gelatine, fruits, and salad moulds are used as color and form contrasts for the speciality dishes of decorated hams, turkeys, and fish. Gloss is achieved through the use of sweet or savory aspic glazes. For fruit, white glaze can contain lemon and sugar; for meats, the glaze may include brown oxtail soup and Madeira for fish, it can be colored green with dill. Food color may be used to make aspic frosting for decorating rims of glasses. Thin slices of white radish, raw and cured fish can add the dimension of translucency. Opacity and diffuse reflectance can be added to the table with folded napkins and paper frills. Colors can be added with moulded and piped marzipan, buttercream, icing, and chocolate.

Plates of butter may be made more attractive with greenery or flowers made from radishes. Gourds and cucumbers are versatile in that they can be grooved in the whole or in slices, cut into wedges, balls, and comets, made into boats and turrets, or into decorated bowls that can be stuffed with mixed vegetable salad. Colorful and shapely mixtures can be made from pimento, tomato, cucumber, and egg. Decorations may be finished off with angelica, cocktail cherries, carrot twists, or grooved lemon baskets. Rarely are all ingredients in a dish perfect. If there is one item that has an especially fine appearance, it should be placed where the eye cannot miss it.

Desserts can look more exciting when served in glasses, and crystal adds a dimension of liveliness to a scene. However, crystal or colored glass prevents the serious wine drinker fully appreciating the proper color of a wine. This is why variety of presentation is added through serving different wines in glasses of different shapes.

Although there are restrictions to the colors appearing on our plate there are few restrictions to shape. We can carve, circle, dice, strip, or bunch vegetables to increase perceived variability and increase the perceived texture and flavor variation. Shapes add an important visual element to decorative platters and individual dishes. Some of the color/shape combinations possible using a number of raw and cooked (or blanched) vegetables are listed in Table 9.1.

The principles outlined in the last section are appropriate for all displays and dishes but the highest form of pictorial coloring to emerge from the kitchen is the iced and decorated cake. This can be an art form in its own right. Included into the overall design and display are color combinations that have become traditional for specific occasions. These are listed in Table 9.2.

It is the third dimension that taxes the imagination most. Buffets are often held in large rooms so it is not sufficient merely to make a two-dimensional spread on a flat surface. For the display to be at its best there must be a bold

TABLE 9.1. Colors and Possible Shapes for Vegetables (Emery, 1980)

Vegetable	Color	Circles	Dice	Strips	Bunch
Raw					
Cabbage	Red			x	
Celeriac	White	x	x	x	
Celery, chicory	White				x
Cress	Green				x
Cucumber	Green	x	x	x	
Curly endive, lettuce	Green			x	x
Olives	Black, green	x	x		
Onions, spring	Green	x	x	x	x
Parsley, watercress	Green			x	x
Peppers	Red, green, yellow	x	x	x	
Radish, tomatoes	Red	x	x		
Cooked or blanched					
Artichoke bottom	Green		x	x	
Asparagus, broccoli	Green			x	x
Aubergine	Mauve	x	x	x	
Carrots	Orange	x	x	x	
Cauliflower	White				x
Courgette	Green	x	x	x	
French beans	Green		x	x	x
Leeks, potato, parsnip	White	x	x	x	
Mushrooms	Brown	x	x		
Peas	Green	x			
Sprouts	Green				x

TABLE 9.2. Colors Associated with Special
Occasions (Emery, 1980)

Occasion	Colors
New Year's Day	White, apple green contrast
St. Valentine's Day	White, deep red contrast
St. Patrick's Day	White, emerald green contrast
First day of spring	Yellow, coral, or apple green
Easter	Yellow and chocolate brown
April Fool's Day	Pale yellow, green contrast
Mother's Day	White, pink contrast
May Day	Pastell shades contrasting
First day of summer	Pastels of the spectrum—not blue
Hallowe'en	Orange and dark brown
First day of winter	White, dark green contrast
Christmas	White, red, and green

excursion into the height dimension. This announces the whereabouts of the food, thereby encouraging those present to eat. It also forms a focus for what may otherwise be a featureless room and projects an image of the care taken to prepare the meal. Height contrast is achieved with baskets of bread, flowers, candles, and sockels (large sculptures of bread, ice, butter, fat, chocolate, cake, or rice). Beautiful Thai sculptures fashioned from fruit and vegetables, and flowers made from soap can be used to create an artistic center for the display. Colorful bowls, trays, and shining silver lend an appearance of elegance. The buffet is a showcase for the chef so if there is a skilled cook/designer/artist on the crew there is scope for an eye-catching display.

An overall display image of cleanness, freshness and artistry creates a feeling of trust in the cook as well as a good appetite. Foods exhibited are selected to last the length of the meal. A melting ice sculpture adds a desirable dynamism to the display. The dynamism produced by curled sandwiches and drying slices of meat is not so welcome. The highest quality food display is one that, in addition to the food, contains extremes of opacity and whiteness (as in linen), extremes of transparency and gloss, and extreme cleanness.

On the Plate

During World War II, when foods were rationed and scarce, the British Restaurant was noted for its gray-colored meals of thickened soup, mince with gray beans and potatoes, followed by thin apple stew. Today an abundance of fresh fruits and vegetables coupled with imported exotic fruits fill our plates with beautifully colored combinations.

In the compilation of a menu, clashes of color and repetition ought to be avoided. Interest is added if, from course to course, a basic color sequence incorporates variation and contrast. The colors present in a single serving ought not to be too insipid or too bold, with the emphasis on natural rather than manufactured color. If the color of the main element of the course is rather heavy, pastel shades should be aimed at for the garnish, or vice versa.

Color contrast is needed to avoid monotony and create interest. The contrast phenomena described in Chapter 2 are applicable to food presented on the plate. Similar colors tend to detract from each other, while complementary colors tend to reinforce. For example, red presented on a red background looks less red, while green broccoli will enhance the redness of red fish. Purple (blue plus red) grapes look less red on a red background and more red on a blue ground, while orange will look less red on a red ground but redder on a yellow one. Bright saturated colors of defined shape (an orange slice, carrot slices, a sprig of green parsley) can be used to accent less pure, duller colors. Shape and texture contrasts can be used to alleviate impressions of blandness. For example, broccoli or peas will set off poached white fish dishes better than spinach or mashed carrots. Accent colors of bright orange, red, or green vegetables give color interest to the duller tones of meat and roast potatoes. Colors create quality expectations; unhealthy looking fresh greens and overcooked degraded vegetable greens lead to a rapid downgrading of meal quality expectations.

Color and form vary naturally on the plate according to the meal displayed. However, this combination can be further increased by inclusion of herbs, garnishes, and flowers. Their purpose is not merely to add color and form but to increase sensations of scent and flavor. Skilful choices of flowers enhance and complement particular tastes and flavors. The shape and color of a loaf of bread is characteristic of its type and flavor, particularly in Europe.

There are also opportunities for contrast in translucency. Cloudiness of consommé, which should be perfectly clear, results from the use of poor quality or greasy stock. At the other end of the soup opacity scale is cream soup that need not contain cream. "Cream" refers to translucency and consistency rather than the content. Transparency is brought into a meal through inclusion of consommé, some sauces and gelatine desserts. Translucent products may be obtained with gelatine or jellies perhaps mixed with a little milk to introduce clouding. Aspic gels are used to create gloss as well as clarity on the plate. The rare but burned steak adds its own type of gloss.

Appearance contrasts induce the anticipation of contrasts of in-mouth texture, both forming part of the enjoyment of eating. The combining of textures is as skilful as the combining of appearance contrasts. However, there are different effects of age on the appreciation of color and texture contrasts. Younger people tend to appreciate a variety of colors, but not of texture. With maturity comes an increase in appreciation of texture contrasts and a lessening of a demand for a profusion of colors. Over-decoration and over-elaboration is distasteful to most adults, but not to children. In the Western dining room the most satisfactory impression is usually created when the food, comprising sufficient color and shape contrast, is set onto the plate in a simple way. Over elaboration with too much garnish or too many items on the plate draws attention away from the main element of the dish.

The appearance and presentation of food on the plate is a matter of pride to the conscientious waiter. If there is a crest on the plate, this must appear at the top, the side away from the diner. If the part of the plate nearest the eater is designated 6 o'clock, the main element of the dish is placed between 4 and 8 o'clock. More control over the cutting of this element, often meat, can be achieved if it is placed here. Garnish such as cress or watercress may be added next, nearer to the center of the plate. Vegetables are then added between 9 and 12, and 12 and 4 o'clock, leaving room for the potatoes. Sauces and gravies are finally added where they belong or to the diner's instructions. A symmetrical effect is strived for, using as far as possible the natural shapes of each element to fit into the curve of the plate. These principles obviously do not apply to all cuisines. Placement is merely the concern of the individual during, for example, a Chinese or Indian subcontinent meal in which it is the custom to serve oneself from centrally-placed dishes.

Similar principles are followed for the serving of desserts. A slice of gateau should be presented on a sweet plate with the pointed end placed directly in front of the diner. From this angle the diner may inspect the visual quality of the filling, and both the contrast inside and outside texture and structure at the same time.

Both serious and casual eaters prefer food images to be visually positive. Whether we live to eat or eat to live we demand a meal that looks good; color and appearance must tempt. This is especially the case for those who are sick. Illness has psychological repercussions affecting attitudes, often taking the form of pickiness and finickiness. Older people tend to be more health conscious, and their olfactory and taste sensitivities decline with age. Although vision sensitivity also declines, older members of the population are comparatively more sensitive to color cues and less sensitive to changes in flavor concentration. As they also perceive colors as having less chromatic content than those who are younger, there is a case to be made for meals of greater coloredness to be served to older people in hospitals. Meals for the sick should be attractive, appealing, and appetising, even to the extent of sacrificing nutrition for short-stay patients. Ways of increasing the acceptability of hospital food include reducing the frequency of similar foods on the menu, providing choices among the entrées and varying food combinations, providing menus with more description of the items, and by providing attractive trays. Most or all of these items are high in visual content.

FOOD COLOR IN TRADITION

Fashion, tradition, and folklore play a great part in our food ways but they change with time and differ with geography and ceremony. What was a popular item on the menu may after a few years gradually disappear from view, as it becomes less fashionable. Also, geographic availability of natural species and ingredients inevitably leads to local preferences. In addition different foods are associated with different calendar customs and family events.

Food Symbolism and Fashion

There are four types of association or symbolism associated with architecture (Smith, 1980), but they are also appropriate to foods. These are associational, acculturated, symbolism of the familiar, and archetypal symbolisms. *Associa-*

tional symbolism concerns personal experience that arises from upbringing and education. Color forms a major part of this symbolism in the marketplace. For example, white is associated with dairy products, softer brown and golden tints with expensive luxury foods. Similarly, colors are used to identify specific locations in some stores. On a cold day, we feel warm at the sight of a hot meal, but the same material can change in association according to current environmental conditions. The sight of a meal may change from being welcome to being not needed when an unwelcome visitor arrives. *Acculturated symbolism* concerns cultural influences. For example, roast beef and Yorkshire pudding are the essence of Englishness. Green foods represent a new beginning at Jewish New Year celebrations, round things such as onions, ring-shaped pastries, and meat balls signify continuity and the hope the year will be rounded, gold items such as sliced carrots, pumpkin, squash, and honey represent prosperity, and chickpeas and beans represent abundance. The dark and black of aubergines, olives, or chocolate is nowhere to be seen. Food itself is symbolic and can be accepted as such without regard to its nutritional value. Some religious foundations specify rigid food acceptances or taboos. *Archetypal symbolism* is the symbolism of psychologists. It is that which may lie in the subconscious part of our personality. Red and yellow colors are called *warm*, blue and green *cold*. *Symbolism of the familiar* concerns the routine of everyday life. Everyday foods of correct appearance eaten at normal meal times represent security. An incorrectly or unnaturally colored food arouses suspicion and is unsettling. Fashionable foods become, for a time, part of our symbolism of the familiar.

Of the wonderful dishes created by the chef, some trickle down to the remainder of the eating community, becoming for a limited period a fashionable item on the menu if they can be made economically. During the first half of the twentieth century, our income structure and class system dictated that the great majority of the population was unlikely to have tasted more exotic species of fish such as turbot, and fish paste was eaten in imitation of foie gras. The restaurant trade was able to flourish because of the wide provision of French-style cookery education. This in turn led during the 1980s and 1990s to the growth of a nouvelle English cuisine style. This generally was a high quality, but expensive style, which attracted the affluent and, because of the minute portions, the health conscious.

Salad bars became fashionable when the perennial quest for weight loss, fuelled by the cult of the thin on the screen and in the press, gained momentum. They continued through the cult of the fit and the gym and subsequently because of the "bad for us" high-fat diets. "Good for us" is a mixture of colored fruit and vegetables eaten every day. Above all it is the vivid colors bringing varied textures and flavors that draw us to the salad bar, especially during warm weather. In Britain, British adaptations of Indian or Chinese cuisine continue to gain ground, and on the high street take-aways serving such cuisine are found more frequently than the traditional fish and chip shop. Japanese and other ethnic foods have been slower to gain acceptance. Cajun cooking has thrived in Louisiana for generations, but after hitting Britain briefly, declined. Use of poor quality spice mixes contributed to the fall. Mutton is no longer available in Britain and lamb has taken its place.

A perceived or advertised benefit of a particular food item may lead to an increase in liking through word of mouth, imitation, then habit. We may then receive a counter impulse by the publicity of a food scare. Many Britons once went to bed with a cup of Horlicks or Ovaltine until a downgrading in the reputation of milk fat contributed to a decline in consumption. The bad news about fat and frying has led to the conversion of the once traditional and commonplace fried breakfast into a luxury eaten occasionally for a treat or when away on holiday. Other health news has led to the demise, except to the adventurous, of the raw egg and steak tartar (raw beef).

Among the brands that have come and gone since the 1950s are many types of breakfast cereals, the open sandwich, and impulse products such as icecreams, confectionery, and the synthetic comfort food Angel Delight. Products that have come and stayed are frozen vegetables, fresh fruit and vegetables, and the blackcurrant drink Ribena. Heinz baked beans, Cadbury's milk chocolate, and Coca-Cola have been around even longer and are still heavy sellers. We can now prepare many cuisines at home. A full range of pastas and sauces for them has arrived to supplement canned spaghetti in tomato sauce. Ranges of rices, curry sauces and powders, with noodles and Chinese ingredients allow us to present the world of cookery to our family and friends.

The French education led to a change of menu vocabulary. For example, words such as *chiffonade* replaced *chopped* and *shredded*. Vegetable cuts were varied and named, *julienne* are fine strips with 2-mm cross-section and 40-mm length, *jardinière* are batons of 3-mm cross-section and 20-mm length, *macè-doine* are 6-mm dice, *brunoise*, 3-mm dice, *paysanne* described regular geometric cuts such as squares, triangles, and circles, while *losange* are diamond shaped. Among French color terms on the menu are *au bleu*, used to describe very underdone meat, *vin blanc* and *vin rouge* in reference to wines, while *la blanquette* is a white stew, *doré* is golden, *clair* is clear, *gratiner* is to color in a hot oven using grated cheese or breadcrumbs, *rissolé* is to fry golden brown, *tomaté* is to color using tomatoes, and *vert* is green. The appearance of the menu in Britain has generally evolved from the sole use of the French language to a sensible combination of English using some specific French preparation and cooking terms.

All these practices and feelings are culturally, family, or just individually based. Each family has its own food routine behavior pattern and its own individualities and traditions of food selection at times of celebration, happiness, and sadness.

Traditional Foods

Food taste and appearance are normally regionally based through the availability of local produce, and shaped by economic and political events. In North America, for example, all culinary regions are traceable to the specific populations of American Indians, pioneers, missionaries, cowboys, and immigrants. Oysters provide an example of preference linked to availability. These bivalves sometimes develop green gills. In areas like New Brunswick, where this condition is uncommon, they can be rejected, but in regions such as Virginia, where they are more common, they tend to be preferred, and in Marennes in France they are highly prized. Another example of local preference is linked to the color of tortillas. This derives from the color of the maize used in its production; in Mexico they are turquoise, in the USA, bluish gray.

Even in a small country such as England there are regional food differences based on color. Darker foods are preferred in the north. Treacle toffee and liquorice are preferred sweets, and icing sugar is dyed colors that are darker and more saturated. Proportionally higher sales of darker beers, brown sherry, and Demerara rum occur there. In the southwest counties of Devon and Cornwall there is a preference for white in foods. Fresh milk and cream are not only used in great proportion with their famed cream teas, but on fruit and vegetables also.

There are also racial differences. In North America, individuals with a Mexican food background tend to regard appearance as a less important attribute, while those with a Chinese background regard it more highly. In a supermarket trial in which yellow and white cheeses were displayed, individuals from the black and other ethnic groups tended to select yellow cheeses, white individuals tended to select white. There are also individually-based preferences. Again in the USA, significant differences in preference occur between those who are overweight and underweight, between males and females, and between black and white individuals. Health-conscious panellists give snack foods significantly higher scores for greasiness than tasters who are flavor orientated.

Food names signify different things to different peoples. This is because the name stands not for a physical object but for a concept or a sign. Rice and potatoes are both white in color, but rice can be also ethnically white. In the Andes, a neighbor may be seen as pretentious for cooking it. An expensive food stands for wealth, a cheap one for poverty. In the normal eating situation, foods must be within the color range the eater has been brought up to regard as proper. Although potatoes and potato products may be presented in a great variety of sizes and shapes, their color must be within the correct range. Although we may accept color changes in dress and decor, we are reluctant to accept inappropriately colored food.

In the Western world, *naturalness* is a quality to which we can become conditioned. In Britain white is associated with lack of flavor and white-shelled eggs are associated with battery production. Association of the brown color of brown-shelled eggs may be due to an association with farms and the earth and so with naturalness and of being 'good.' Brown eggs are common in southern New England, but white are preferred in New York and the Middle East. Such preferences may be founded on product quality as brown eggs contain a higher yolk weight and yolk/albumin ratio than tinted or white-shelled eggs. Meat products such as paté score low on naturalness because they have a non-natural appearance. However, we often demand that the natural appearance of meat and fish be disguised or hidden from sight by a crumb coating. Perhaps we have an aversion to eating parts of what were living creatures. Feelings may be different on festive occasions when the natural forms of the sucking pig, whole salmon, or roast turkey may even be enhanced, possibly through a subconscious link with the hunt.

The debate over whether meat ought to be bloody and rare or well done probably has as much to do with appearance and the sight of blood as with texture. In France all meat, except chicken, is preferred rare. The reason for the exception is the high risk of salmonella. Anthropologically, to eat bloody meat is making a potent statement of power over other highly evolved animals. In Britain, rare cooked meat is interpreted by the French as being overcooked. Among Australians there is wide disagreement when ordering beefsteaks.

Existence of migrant workers encourages marketing of foreign foods. For those who feel alienated a sense of identity and reassurance is provided by a diet of their own regional food. This in turn increases the choice available to the indigent population. Similarly, food not eaten since childhood can bring back fond memories of upbringing and security. The back-to-nature trends of cosmetics and toiletries, the successful and lucrative marketing of regional dishes, rural-fresh vegetables, and foods made from old recipes, result in appearance images and feelings of warmth, comfort, and good health. An identification of some North Americans with Ireland, and therefore with green, leads them to coloring the River Chicago green and to drinking green beer annually on St. Patrick's Day.

Food Folklore

Oral tradition and folklore play a large part in our view of foods. Although over a great area of Europe and the East many colors are used to dye eggs, red is by far the most popular. In 2900 BP they were exchanged by the Chinese at spring festivals. Today in Britain they are used in Easter games such as egg rolling. In Romania, where red-painted eggs are called love apples, and Slovenia they represent love and health. In Hungary, where the name for Easter eggs is *piros tojas* or red eggs, and in Russia, Yugoslavia, and Greece they represent blood lost by Christ on the cross.

On every continent rice has long been the staple diet of half the human population. It is not surprising, therefore, that it has become the object of folklore and tradition, much of which concerns the use of color. In China, rice is called *white jade beads*, and the phrase *having a bowl of rice* means having a steady job. Although the most common variety is white, it does occur naturally in a number of other colors. Black rice is used as a tonic for the kidneys, red for modulating one's spirit, and there are also light-yellow and light-violet varieties. Rice dyed red is given to wedding guests. In some provinces, cooked black rice is eaten in the lunar calendar of April to commemorate the birth of an ox, or to mark the struggle against miscellaneous levies. White rice forms one part of the collection of bright colors, which symbolize happiness in ceremonies practised by the Canchis Indians in southern Peru. Rice colored yellow with turmeric or saffron is widely used in custom in India and Pakistan and by Indians in Malaysia. It is used to scare away demons affecting fertility; it forms an invitation to wedding feasts, and is sprinkled on the couple afterwards. White rice is used as a contrast color in celebratory dishes. In Japan, red rice balls (steamed rice mixed with red beans) form an essential part of the celebration of happy occasions such as weddings, New Year's Day, and anniversaries.

Aspects of appearance as well as color are important in traditional foods. On Lammas (loaf mass) Day, celebrations in medieval England included the baking of loaves colored red (with rose-petal), golden orange (saffron), yellow (lemon), green (parsley), blue (thistle), indigo (plum), and purple (violet). Architectural shapes of castles and warships, depictions of animals, Eve in the Garden of Eden, star and moon shapes were part of the traditional fare. There is a fine tradition of shaped breads in central Europe. These are used to reinforce the Christian belief of the transubstantiation to the Body of Christ. Christstollen is a sweet yeast loaf with fruit and nuts swaddled in white icing to represent Jesus. This was first documented in 1329. Traditions involving shaped breads occur in eastern and western Europe, Scandinavia, and Mexico. Ceremonies concerning them range from reinforcement of religious beliefs, to celebrating the arrival of seasons of the year, to increasing fertility of the fields, and to bringing good luck. Pretzels were believed to drive witches away as there is no escape once they become locked in the never-ending loop. Iced fruitcakes, optionally tiered, are customary for weddings and christenings in many countries. Gloss plays a part in food folklore also. The Japanese ceremony of 'breaking the mirror' is carried out when a new wooden cask of sake is broken open with a mallet. The title refers to the fact that the surface of the drink is so clear it looks like glass. People of all religions and races ceremonially share foods during celebrations. Table settings and room decoration also play their part in festive meals.

Dance, song, and ceremony are relevant parts of the food production process. Dances were used to greet the return of the growing season of spring, to accompany sowing, to make rain, to celebrate the end of harvest, and to accompany the stirring of the mincemeat and the Christmas pudding mix. Customs vary around the world, but in no part are all of these traditions and rites obsolete. The cooking of a flambé at the table, the preparation of a pizza or a sushi in front of a restaurant customer are theatrical performances. A well-lit and displayed supermarket bakery department is part of a performance-oriented personal relationship between the store management and the customer.

REFERENCES

Allen, D. E., 1968, *British Tastes*, Hutchinson, London.
Anker, M., and Batta, V. K., 1987, *Basic Restaurant Theory and Practice*, Longman, Harlow.
Anon., 1997, *Description of Food and Drink on Menus*, leaflet published by Warwickshire County Council Trading Standards Department.
Aron, J.-P., 1975, *The Art of Eating in France*, translated by Nina Rootes, Peter Owen, London.
Biller, R., 1986, *Garnishing and Decoration*, Virtue, Coulsden.
Birnbaum, D., Steiner, J. E., Karmell, F., and Ilsar, M., 1974, Effect of visual stimuli on human salivary secretion, *Psychophysiology* **11**: 288–293.
Bode, W. K. H., and Leto, M. J., 1984, *The Classical Preparation and Presentation of Food*, Batsford, London.
Brown, P. B., and Day, I., 1997, *Pleasures of the Table*, York Civic Trust, York.
Cousminer, J., and Hartman, G., 1996, Understanding America's regional taste preference, *Food Tech.*, **50(7)**: 73–77.
Cracknell, H. L., and Nobis, G., 1989, *Mastering Restaurant Service*, MacMillan, London.
Emery, W., 1980, *Culinary Design and Decoration*, Northwood, London.
Hagen, A., 1992, *A Handbook of Anglo-Saxon Foods*, Anglo-Saxon Books, Pinner.
Hammond, P. W., 1993, *Food and Feast in Medieval England*, Alan Sutton, Stroud.
Hieatt, C. B., and Butler, S., 1976, *Pleyn delit*, University of Toronto, Toronto.
Hope, A., 1990, *Londoners' Larder*, Mainstream Publishing, Edinburgh.
Hutchings, J. B., 1998, Colors in plants, animals and man, in: *Color for Science, Art and Technology*, K. Nassau, Ed., Elsevier, Amsterdam, 223–246.
Hutchings, J. B., 1999, *Food Color and Appearance*, 2nd edn., Aspen, Gaithersburg, MD.
Hutchings, J. B., Akita, M., Yoshida, N., and Twilley, G., 1996, *Color in Folklore with Particular Reference to Japan, Britain and Rice*, The Folklore Society, London.
Le Cordon Blue, 1994, *Classic French Cookbook*, Dorling Kindersley, London.
Lyman, B., 1989, *A Psychology of Food, More than a Matter of Taste*, van Nostrand Reinhold, New York.
Mead, W. P., 1967, *The English Medieval Feast*, Allen and Unwin, London.
Mills, I. J., 1989, *Tabletop Presentations*, van Nostrand Reinhold, New York.
Moskowitz, H. R., 1983, *Product Testing and Sensory Analysis of Foods*, Food and Nutrition Press, Westport, CT.

Newall, V. J., 1991, Colour in traditional Easter egg decoration, in *Colour and Appearance in Folklore*, J. B. Hutchings and J. Wood, Eds., The Folklore Society, London, 36–39.

Noel, P., and Goutedonega, R., 1985, Saucisses thermques de type francfort a base de proteins vegetales, in *Proceedings of the 31st Eur Meet Meat Res Work, Albena, Bulgaria*, 786–789.

Radhakrishnan, K. K., 1980, Colour and textile design, in: *Colour 80 Seminar Proceedings, A Colourage Supplement*, 43–48.

Rolls, B. J., Rolls, E. T., and Rowe, E. A., 1982, The influence of variety on human food selection and intake, in: *The Psychology of Human Food Selection*, L. M. Barker, Ed., Avi Publishing, Westport, CT, 101–125.

Roux, M., 2000, *Life is a Menu*, Constable, London.

Schutz, H. G., and Wahl, O. L., 1981, Consumer perception of the relative importance of appearance flavor and texture to food acceptance, in: *Criteria of Food Acceptance Symposium Proceedings*, J. Solms and R. L. Hall, Eds., Forster, Zurich, 97–116.

Smith, P. F., 1980, *The Dynamics of Urbanism*, Hutchinson Educational, London.

Steiner, J. E., Konijn, A. M., and Ackermann, R., 1977, Human salivary secretion in response to food-related stimuli, *Israel J. Med. Sci.*, **13**: 545–553.

Symons, M., 1982, *One Continuous Picnic*, Duck Press, Adelaide.

Tannahill, R., 1988, *Food in History*, Penguin, London.

Weismantel, M. J., 1988, *Food, Gender and Poverty in the Equadorian Andes*, University of Pennsylvania Press, Philadelphia.

Wheeler, A., 1986, *Display by Design*, Cornwall, London.

Index

NOTE: All venues, such as cafes, pubs and restaurants, at which food and drink is available for consumption on the premises, are normally indexed under Restaurant.